SCIENCE

& the Search for Meaning

SCIENCE
& the Search for Meaning

ॐ

Perspectives from International Scientists

EDITED BY JEAN STAUNE

FOREWORD BY PHILIP CLAYTON

Templeton Foundation Press
Philadelphia and London

Templeton Foundation Press
300 Conshohocken State Road, Suite 670
West Conshohocken, PA 19428
www.templetonpress.org

Introduction translated by Caroline Rossiter; chapter 1, chapter 8, and chapter
13 translated by Tom Mackenzie; chapter 9 translated by Caroline West

Christian De Duve, "Mysteries of Life: Is there 'Something Else?'" is a slightly
revised version of the original printed in *Perspectives in Biology and Medicine*
45, no. 1 (Winter 2002): 1–15. © The Johns Hopkins University Press.
Reprinted with permission of The Johns Hopkins University Press.

Michael Heller, "Science and Transcendence," *Studies in Science and Theology*
4 (1996): 3–12, reprinted with permission from the European Society for the
Study of Science and Theology.

*Templeton Foundation Press helps intellectual leaders and others learn about
science research on aspects of realities, invisible and intangible. Spiritual realities
include unlimited love, accelerating creativity, worship, and the benefits of
purpose in persons and in the cosmos.*

Designed and typeset by Kachergis Book Design

LIBRARY OF CONGRESS CATALOGING-IN-PUBLICATION DATA
Science et quête de sens. English
Science and the search for meaning : perspectives from international scien-
tists / edited by Jean Staune.
p. cm.
Includes bibliographical references and index.
ISBN-13: 978-1-59947-102-0 (pbk. : alk. paper)
ISBN-10: 1-59947-102-7 (pbk. : alk. paper)   1. Science—Philosophy.
2. Science—Methodology.   I. Staune, Jean.   II. Title.
Q175.S359 2006
501—dc22
2006010778

Printed in the United States of America

06 07 08 09 10 11   10 9 8 7 6 5 4 3 2 1

Contents

Foreword

PHILIP CLAYTON

Even those who most disagree with the tenor of this book will have to acknowledge that it is an exciting time to be reading and reflecting on the implications of the sciences. Only in a few periods of the history of modern science—the Renaissance and the birth of modern science, the early responses to Galileo and Newton, the heated responses evoked by Darwin, and the early reactions to relativity theory and quantum physics—has there been such a clear opening for connecting science and the transcendent. And at *no* other point in the history of modern science have so many distinct debates converged upon a few central questions:

- Is the world studied by science the only reality, or does it point to a deeper reality?
- Is nature a random and chance process, or a project with a purpose?
- Can humanity be fully understood in terms of the natural sciences, or is there a transcendent dimension to human existence?

I

It was certainly unexpected that the period of the greatest breakthroughs in the history of science would also expose the greatest limitations on the scope of scientific knowledge. Is it not ironic that the best verified equation of motion in the history of physics, the Schrödinger wave equation, would be connected with an inherent limit on knowledge of the quantum world? And was it not surprising to learn, just as we completed mapping the human genome in the Human Genome Project, that the dream of genetic reductionism was impossible because there are too few genes (only about 30,000) to code for all but a few human behaviors? It is not remarkable that humanity would come to have such good scientific reasons to know that there are things we will never know: the exact location and momentum of a subatomic particle at some moment of time; the future states of a "chaotic" system, given that its present state can never be measured with sufficient accuracy; or the state of the universe outside our "light cone" or before the big bang.

Yet none of these limitations, and the many others described in this book, show or imply that the project of science is itself bankrupt. The limits are humbling to our desire for complete knowledge, yet they are not mortal wounds to the human quest to know our world by scientific means. Of course, a different result was possible. We might have learned that the scientific project is fundamentally flawed, that the quest for scientific explanations is absurd because nature is not lawlike or because all systems are as unpredictable as chaotic systems. But this has not happened. Indeed, the fact that science is powerful enough to demonstrate its own limitations is a good reminder of what an effective means of knowledge it is. Nonetheless, it has turned out that science, that potent aid to human knowledge, is not all powerful. Science, we now know, can tell part of the story, but it cannot tell the entire story.

An interesting example of this new "yes and no" to science is the discussion of emergence in the natural world.[1] Scientists have recently begun to understand how, as nature increases in complexity, new levels of organization emerge—the biological out of the chemical, the psychological out of the neurophysiological.[2] The biophysicist H. Morowitz has even identified twenty-eight distinct levels of emergence in natural history.[3] On the one hand, the new phenomena that appear over the course of evolution remain dependent on earlier levels of universal history and thus on the biological, chemical, and physical laws that govern those levels. For example, consciousness cannot be fully understood without understanding the nature of the human brain and the history of its evolution; the same is true for all other emergent phenomena in the evolution of the cosmos. On the other hand, the newly emergent phenomena cannot be fully understood in terms of the lower-level laws on which they remain dependent. For the evolutionary process continually produces new sorts of systems, with new types of entities and causal processes. Hence, a full understanding of the new levels requires explanations *given in terms of the emergent phenomena themselves*. The new theory teaches that emergent phenomena are irreducible with regard to their causes, their explanations, and hence their true nature as objects or processes.

What is true for emergent phenomena is true also for comprehending the directionality of the process itself; no explanation at a "lower" level can explain why the process would eventually produce the higher-order phenomena that it has produced. Explaining the process as a whole requires a theoretical perspective broad enough to include the "highest" point reached by the process so far. Indeed, since the process of evolution continues, we suppose that a higher standpoint is needed than any that nature has reached so far. This was also the position taken by Teilhard de Chardin.[4] (Of course, one can accept emergence theory without claiming the degree of knowledge of the future that Chardin claimed.)

Each of the authors in this book responds in a different way to the new evidence that reveals the limitations on scientific knowledge. One can distinguish three groups of authors. The first group advances clearly religious positions on the nature of the "other reality" that transcends scientific reality; from that perspective, they are able to speak of what science knows, what science can never know, and what another kind of knowing might look like. The more cautious authors, in the second group, still affirm that science cannot explain all parts of our experience. But their arguments are more analogous to the classical *via negativa*, insofar as they point toward a deeper reality, a veiled reality, which relativizes the reality known to science but that (they argue) never gives itself to us to be known.

A third group of authors stands between the other two, although here there are sharper differences among the various authors. For these authors, science provides us with at least some knowledge of the *Jenseits*, some hint of what lies beyond. They argue that science—or science supplemented by philosophy or morality or poetry—does not merely declare its own limits; it also begins to indicate the nature of the transcendent. Some signs within the natural world, which Peter Berger calls "signals of transcendence,"[5] open a window that allows us to see vistas of another realm altogether. "Now we see but a poor reflection as in a mirror" (1 Corinthians 13:12, NIV), yet we do see *something* of what lies beyond. Beyond this minimal point of agreement, however, speculations differ. Some of the authors believe in a reality that transcends the natural order altogether, whereas others discern a deeper level that grounds or produces all natural realities. Nevertheless, the thinkers in this third group are agreed that the natural world, when studied carefully, gives signs that there is more to know than what the natural sciences can reveal to us. And it is science that gives us the first hints of what this "something more" is and how it can be known. Perhaps one hears here the spirit of Pascal:

Man is only a reed, more frail than nature, but he is a thinking reed. It does not need the whole universe to wipe him out; a breath, a drop of water, is enough to kill him. But when the universe wipes him out, man will still be more noble than what kills him, since he knows that he dies and knows the advantage the universe has over him. The universe knows nothing. . . . It is not at all in space that I must seek my dignity, but in the ordering of my thought. I would have no advantage at all in possessing the earth. By space the universe embraces me and swallows me up like a point, but by thought I understand it.[6]

II

The wise man is the one who knows which opinions can be altered by the force of the better argument, which opinions should be altered but will not be, and which opinions go beyond matters of argumentation altogether. Bernard d'Espagnat maintains that the choice between his two major theories of the Real falls in the third category. Yet there is another distinction among the authors of this book that is equally fundamental and which may *precede* rational debate rather than respond to it.

One detects a certain cautious or skeptical attitude in the writing of some authors regarding science and the beyond, and a certain boldness in the responses of others to this topic. Certainly, both groups are represented in this volume, and the reader needs no help from a foreword to distinguish between the two. (Indeed, it would seem that the foreword and the introduction of this book may have been divided between authors representing the two types!) In history it is usually the bold thinkers who have introduced the major new paradigms of thought. These thinkers are quicker to see the tentative implications of their field of study and to follow these implications outward into new uncharted territory. The bold authors are quicker to argue for the validity of other kinds of knowing. They look for plausible connections and grand coherence, and they are more likely to insist, "How will we know whether the new paradigm is plausible unless we first explore it?" By contrast, the cautious or skeptically minded thinker is an expert at the suspension of belief, at balanced agnosticism, at the *epoché* of Husserl. Perhaps the eyes of such a thinker are equally skilled at seeing the possible implications of both the knowledge and the limits of science. But he or she believes that it is wiser to describe many possible connections, many possible implications, than to select just one theory of ultimate reality as true.

As I said, both types of thinkers are represented in this book. The bold authors see in the ordered world of physics a sign of a Creator who has ordered it; they see in the broad patterns of biological development an indication of purpose in nature, and they see in consciousness a proof that humanity will only be understood when we include the spiritual dimension in our explanations. The cautious or skeptical thinkers encourage their readers to pay attention to each of these possibilities, to keep an open mind, to wonder whether the world may not be massively more complex, more elusive, and more mysterious than we have supposed. But where the bold thinker sees proof, or at least scientific evidence, the cautious thinker sees grounds for speculation and no more. Where the bold thinker discovers a new metaphysical paradigm, the cautious thinker finds reason to acknowledge limitations in existing paradigms. Where the bold thinker is kataphatic, the cautious thinker is apophatic.

One finds exactly this same distinction in the styles used by the various authors as they discuss the limitations on naturalism. All of the authors in this book appear to reject materialism in the traditional sense of the word, the sense that has been dominant in scientific circles for many decades if not centuries. All the thinkers affirm that there seems to be more to reality than what the natural sciences have presented and are able to present. But beyond this point their responses vary. Some of the authors argue that science has now presented us with conclusive grounds for recognizing the falseness of naturalistic assumptions. By contrast, the cautious thinkers conclude only that the assumption of naturalism is always hypothetical or methodological, for although scientific naturalism is our best means for attaining rigorous knowledge, it cannot comprehend everything that reality is. Reality is grander than any narrow naturalism will allow, even though we may not have the epistemic faculties to comprehend it in all of its splendor. To the bolder thinkers, this reticence is unnecessarily cautious. "You have before you good reasons to conceive reality according to a new paradigm," they respond, "and yet all you will talk about is *what we do not know*. But this is a mistake, for not to know something scientifically does not prove that it *cannot* be known." And, they might be tempted to add, quoting Augustine, "The heart has its reasons that reason knows not of."

The wise person knows which disagreements are fundamental or personal, as d'Espagnat writes—and I suggest that the difference I have just described is one of them. For each reader will likely find himself or herself falling into one or the other of the two groups, and no argument is likely to shift a person from the one to the other. What for one person is evidence that the entire natural world is surrounded by, or enveloped in, or revelatory of the divine, is for the other person merely a hint that there is more in the heavens and on Earth than your theories will ever contain. The fact that I am not disturbed by this disagreement, even if the ambiguity is *never* resolved, is perhaps evidence of which of the two camps I belong in. It seems to me that the two sides represented in this book are in the end allies in helping to undercut all claims for the sufficiency of scientific reason as a means for providing the full range of knowledge that humans need and long for. From this perspective, at any rate, the authors in this book speak with one voice.

What is true of the question of naturalism is also true of the question of meaning. After reading this book, even the cautious reader must conclude that the human quest for meaning transcends any answer that the natural sciences can provide. For it is the essential nature of consciousness to be always *darüber hinaus*, to be always asking why in the face of any statement of fact. Whenever human thought becomes metaphysical—and it does so frequently—it is inevitably characterized by "thirdness" (C. S. Peirce), by the faculty of "synthesis" (Hegel), or by the ceaseless activity of "noesis" (Husserl). The sum total of scientific facts gives us Spinoza's *natura naturata*, the objective side of nature; but it can never give us his *natura naturans*, the underlying source of its becoming—much less *nous noetikos*, the divine "thought thinking itself" in the sense of Aristotle.

Yet we will have made full sense of the world only when we have come to understand not only the totality of facts but also our own drive to make the world make sense. It is no small task. As the great existentialist philosophers of the French tradition have shown, the quest for sense is nothing less than the quest to understand the nature of the human being who poses this question. If the quest for meaning were not fundamental to human existence, humanity would abandon it. But all evidence suggests that we are unable to do so. If, therefore, the quest for meaning is fundamental to our very being, this implies that it cannot be reduced downward to some explanation at a lower level, for to reduce downward is to explain away.

It may be that there is an answer to the human quest for meaning, a Being or a realm that is the answer to life's deepest questions. Many of the authors in this book have presented this belief in a beautiful fashion, and I cannot provide any stronger reasons for this belief than they have already given. I wish instead to make a different point, a point that even the cautious thinkers can accept. A world of materialism, of chance, and of reduction to physicalist explanation can never answer the question of meaning because it lacks the resources even to formulate the question. Only when we give up the goal of reduction, as the results of science are now suggesting that we should, only then can we begin to address the question of meaning and its possible answers. Finally—and this is perhaps the main point—to give up the philosophies of materialism and chance is *already* to have discovered the first part of the answer. For some readers this step will be too little. But, the cautious among us insist, it is *not nothing*.

III

The strategy that I have applied to both naturalism and the question of meaning may at first appear insufficient for the religious or spiritual question. After all, does not religion require knowledge of a supernatural source, of a cosmic purpose, of a transcendent being? Nevertheless, the same strategy is helpful for this question as well. In the discussion with the sciences there may be room not only for bold religious belief but also for a more cautious religiosity.

For many persons, religion is of value only if it offers robust knowledge of the origin of the universe and of its final destination, of the purpose of our life on Earth and the nature of the life eternal. But there are also dangers with claiming to know too much. We see these dangers in the violent form that religious fundamentalism sometimes takes. But the less extreme manifestations of claiming to know too much are also dangerous, as one can see in the present policies of the government of the United States. The overly simplistic religious claims that seem to dominate popular religion in the United States today come to be expressed in international policies and in a warlike attitude, especially toward the Islamic world. Americans (and others) need to learn less boldness and more caution in matters religious. After all, if humans are the most complex organism we know, and if the religious dimension of humanity is connected with our most complex personal and cultural behaviors, must not religion belong among the most subtle, most comprehensive, and most ambiguous expressions of the human spirit? Perhaps the dogmatic claims, the distrust of science, and the intolerance toward other religious traditions that one finds in some popular religions are *more* distant from the true religious impulse than is the caution of those who listen carefully to the methods and results of the sciences but make fewer truth claims.

These considerations suggest the possibility of a vital synthesis of the scientific quest with the religious or spiritual quest. In the end, this is the intriguing possibility that this book most strongly supports. Even the most cautious scientist must acknowledge that there are inherent limits on what can be known by the scientific method. At the same time, as many of the authors here have argued, there are signs that the phenomenal world studied by science is the manifestation of a deeper reality of some sort. Perhaps humanity can only know that other reality through intuition, through speculation, or through a "leap of faith" (Kierkegaard), or perhaps we also possess epistemic faculties that allow for real knowledge of a noumenal realm. That question I must leave open here. It is nonetheless important to recognize that this book offers not one but two different "new paradigms" for responding to this insight. The first paradigm finds evidence within the sciences—both in that which they know and in that which they cannot know—that points to another realm and to another kind of knowledge; and it describes the means, be they faith or intuition or the sense of moral obligation, for pursuing that knowledge.

But the book also offers a second paradigm. It is the paradigm for a type of religiosity that corresponds to the caution of the scientific method and mindset. Of course, this paradigm too must endorse a speculative moment, for there is no religion that is based on algorithms, logical deductions, and scientific inference alone. Nevertheless this second paradigm seeks to walk the religious way with a sort of devout uncertainty, a holy agnosticism, a mystical unknowing. The "scientifically religious" acknowledge that lines of implication move outward from what the sciences know (and from what they cannot know) in the direction of the divine. However, according to such persons, these speculative lines eventually disappear into the clouds that obscure the ontological heights, as the ski lift up the side of Mont Blanc disappears into the grey clouds on a winter's day. Of course, if one then turns his back on the mountain or always remains on the safe ground below, his response will not be a religious response. But some persons, as they begin the ascent, speak of the mountains that rise above them with mystical and apophatic language, being uncertain of what lies above but certain that it is grand and always greater than they can understand.

At one time science was famous for the doors it had closed, the kingdoms it had abolished, the religious claims it had disproved. The present book will help dispel the myth of science as the Great Defeater of all things mystical. Today we instead encounter a science that opens windows onto a rich and mysterious reality. Perhaps we disagree on how much of that reality can be seen and how much will always be obscured by the mists of human ignorance. But we do agree that science does not abolish the human quest for meaning. We agree on the great importance of the new rapprochement between science, on the one hand, and the profound ontological and axiological questions, on the other. The human quest for meaning cannot be pursued in isolation from the sciences of today, even though science alone will never provide the answer.

NOTES

1. See P. Clayton, *Mind and Emergence: From Quantum to Consciousness* (Oxford: Oxford University Press, 2004); P. Davies and P. Clayton, eds., *The Reemergence of Emergence* (Oxford: Oxford University Press, 2006); B. Pullman, ed., *The Emergence of Complexity in Mathematics, Physics, Chemistry, and Biology*, Pontificiae Academiae Scientiarum scripta varia 89 (Rome: Pontifical Academy of Sciences, 1996).

2. See the very helpful treatment of the different levels of reality in the chapter by Thierry Magnin and the discussion of the emergence of the universe by Khalil Chamcham.

3. H. Morowitz, *The Emergence of Everything: How the World Began Complex* (Oxford: Oxford University Press, 2002).

4. P. Teilhard de Chardin, *L'avenir de l'homme*, Oeuvres, vol. 5 (Paris: Editions du Seuil, 1959); idem, *Le phénomène humain* (Paris: Editions du Seuil, 1955, 1970).

5. Peter Berger, *A Rumor of Angels* (Harmondsworth, England: Penguin Books, 1971), e.g., "By signals of transcendence I mean phenomena that are to be found within the domain of our 'natural' reality but that appear to point beyond that reality. In other words, I am not using transcendence here in a technical philosophical sense but, literally, as the transcending of the normal, everyday world that I earlier identified with the notion of the 'supernatural'" (70).

6. B. Pascal, *Pensées*, ed. Philippe Sellier (Paris: Mercure de France, 1976), § 231, § 145.

SCIENCE

& the Search for Meaning

Introduction

JEAN STAUNE

Galileo famously wrote, "Religion tells us how to go to heaven not how the heavens go." In the same way, modernity has been characterized by a kind of philosophical "Yalta": science takes care of the facts while religion takes care of values. In many ways this position may seem perfectly reasonable. It has the advantage of avoiding the confusion between genres and the adverse consequences this can have (e.g., when the interpretation of religious texts leads to the condemnation of certain scientific theories). Moreover, it is a position held by numerous thinkers like Stephen Jay Gould, who popularized it under the NOMA acronym (Non-Overlapping Magesteria).

Nevertheless, the study of the history of Western thought suggests that such "separatism" is no longer tenable when addressing questions pertaining to the meaning of our existence. Did we appear by chance in a universe devoid of meaning? Are we nothing but neuronal beings who no longer need to concern ourselves with matters relating to the spirit? Or is our existence—and the existence of the entire universe—part of a process, or even of a plan? If, on the one hand, science shies away from asking questions of meaning and finality and, on the other, presents a mechanistic, reductionistic vision of man and the world, then surely this famous line of "separation" is transgressed. For in so doing it contributes to the disenchantment of the world, indeed, to the advancement of "non-meaning." Yet scientists blithely transgress the barrier of separation, as demonstrated by the quotations of the following three Nobel Prize winners:

The more we understand the Universe, the more it seems devoid of meaning. —STEVEN WEINBERG[1]

Man can no longer fool himself into believing that he participates in a grand scheme—he knows at last that he is alone in the vast indifference of the Universe, where he emerged by chance. —JACQUES MONOD[2]

The astonishing hypothesis is that you, your joys, your sorrows, your memories and ambitions, your sense of personal identity and free will, are nothing more than the behavior of a vast assembly of nerve cells and their associated molecules. As Lewis Carroll's Alice might have said: 'You are nothing but a packet of neurons.' —FRANCIS CRICK[3]

I would not go so far as to say that these scientists have strayed from their "magisteria." However, it is obvious that scientific theories relating to the origins of the universe, the nature of matter, the nature of consciousness, and the evolution of life must have philosophical and metaphysical implications. These theories cannot be completely neutral with respect to the views that different traditions of humanity have passed down to us about man and the world.

My first point is that separationism is no longer tenable. Without succumbing to the confusion of genres, we have to concede that the outer boundaries of science are not clearly delineated. Moreover, there is an area where science comes into contact and overlaps with areas of religion, spirituality, and the quest for meaning. Philosophical and metaphysical questions should be tackled even if they lie outside the strict framework of science insofar as they are raised by scientific discoveries. A consequence of this first point is that, from the Enlightenment to the twentieth century, science has found itself associated with the spread of meaninglessness, not only for objective reasons (the discoveries of science seem to corroborate the supporters of this viewpoint) but also because the advocates of the presence of meaning in the universe (important Christian scientists, such as Pasteur or Leprince-Ringuet) have been more defensive of the separatist position than their opponents, the defenders of atheism.

My second point is that, during the course of the twentieth century, we have witnessed a complete reversal of trends regarding the metaphysical implications of scientific discoveries. From the study of the infinitely small to the infinitely large, from the study of life to the study of consciousness, a new vision of the world has *already* emerged. This vision "revisits the path to meaning," writes Bernard d'Espagnat, and even goes as far as to allow "the convergence of science and religion," according to Charles Townes. It will deconstruct the mechanistic, reductionist, and materialistic conceptions that characterize the so-called sciences of matter. In the field of cosmology, it will ask questions (without providing answers) about meaning, or even about the existence of a creator. Finally, this vision will show that contingency does not reign supreme in life sciences.

Of course, this is not new. As Sir Arthur Eddington said, "Since 1927 an intelligent man can once again believe in the existence of God!"[4] But it will take decades for such concepts coming from the fundamental sciences to have an impact on the consciousness of the wider public to the point where it modifies its vision of the world (it will most likely take even longer to have an impact on scientists themselves!). The Nobel Laureate Erwin Schrödinger, one of the pillars of the revolution caused by quantum mechanics, declared in 1951, "It will take fifty more years before what we have discovered takes effect on society."[5] That is where we are today. In fact, during the last twenty years or so the public at large, the media, and philosophers have felt that something is happening in science. Dozens of works, conferences, debates, and controversies have drawn our attention to a new scientific paradigm that fully assimilates the question of meaning into this approach, instead of viewing it as an amusing curiosity that we deal with only when we have finished with everything else.

Our present work is situated within this context and is of particular importance for several reasons. The authors are all practicing scientists of note: four are Nobel Laureates and, with the exception of two priests, they express themselves as scientists, not as theologians or philosophers. All of the theories and facts that are referred to have been published in important scientific reviews and form part of an accepted body of scientific knowledge. This does not mean that we reject polemical facts, theories, and people. On the contrary, in a period of changing paradigms like ours, controversies are natural, and it is often the most criticized theories that give rise to major advances. But this work shows—and this is my main point—that it is not necessary to call upon such theories, that the "hard core" of current science provides us the basis that allows science and meaning to come together after a long period of separation.

The authors represent numerous traditions (Islam, Catholicism, the Orthodox Church, Protestantism, Buddhism, Hinduism, and Pantheism) and schools of thought, as I briefly analyze below. They are sometimes contradictory, but what they have in common is that they dispute the validity of certain opinions, like Monrod's, Weinberg's, and Crick's. No, precise scientific knowledge does not lead us to think that we are merely the upshot of random events without meaning. For all the authors of this book, science is no longer introverted but open to the question of meaning.

This work will allow the public to discover the tip of an emerging iceberg. Indeed, most of the texts presented here come from public and private meetings and debates, which the authors have participated in, in the company of numerous other scientists who represent the main scientific disciplines and the main spiritual and religious traditions. Initiated by the Center for Theology and Natural Sciences in Berkeley (www.ctns.org), supported by the John Templeton Foundation (www.templeton.org), and with the participation of the Interdisciplinary University of Paris (www.uip.edu), this endeavor was entitled Science and the Spiritual Quest. Over a period of eight years, this project involved more than 120 scientists and the organization of twenty events in ten different countries (www.ssq.net). At the same time, this work will allow the public to discover the scale, the diversity, the solidity, and the seriousness of this evolution, which promises to lead to a "re-enchantment of the world,"[6] according to the expression of the late Nobel Prize winner for chemistry Ilya Prigogine, who also participated.

This field is rapidly developing, judging by the number of academic chairs and courses being set up to study the theme. But what is the significance of all this? What does it signify to consider the philosophical and metaphysical implications of current science, what we French call *science et quête de sens* (science and the quest for meaning) and the Americans and British refer to as "science and spirituality" or "science and religion"?

My third point will be split into three approaches:

The first approach could be characterized as "apophatic" or "negative," in reference to the theology of the same name, which does not tell us what God is but rather what he is not. It will not tell us anything positive about the question of meaning. On the other hand, it will deconstruct the approaches of those who support non-meaning. Essentially, it is based on negative results, which tell us why we will never know certain things, like the uncertainty principle in quantum mechanics or Gödel's incompleteness theorem in logic.

It is extremely important to understand that this approach is the opposite of "epistemological defeatism," which consists in saying that we will never know this or that or, worse still, drawing the conclusion that this must be where God intervenes (the God of the gaps). Indeed, in these areas, we sometimes have a clear scientific understanding of *why we will never know.* For example, we understand perfectly well, thanks to quantum mechanics, why we will never know the position and the speed of a particle at any one time and, thanks to Gödel's theorem, why we will never have a system of axia that is at once complete and coherent. Bernard d'Espagnat, Thierry Magnin, and Bruno Guiderdoni present this approach, one of the most important results of which, regarding the questions we are dealing with, is the demonstration—once again, scientific, not philosophical—of the non-ontological character of the world in which we live, immersed as we are in time, space, energy, and matter. The fact that another level of reality exists beyond time and space does not in any way prove that a meaning or a project exists in this other level (it could well be completely chaotic). However, it gives renewed credibility to the central affirmations of all of the main traditions (even those that are not monotheistic, like Buddhism or Taoism): the idea that another level of reality exists beyond space and time, and that the human spirit is in one way or another linked to this level. In this way, the doors that classical science had closed are now opening again.

The second approach will be positive (we could make an analogy with cataphatic theology, which refers directly to God). It is about listing the "symptoms of meaning," the facts that, without proof, would seem to suggest that a meaning may well exist in the universe or that our existence is not a contingent event but fits well into a process. This approach is adopted in this book, with more or less force, by Paul Davies, Trinh Xuan Thuan, and William Phillips, who all refer to the anthropic principle, the observation according to which, if we were to change, however minimally, the fundamental constants of the universe, it would no longer be adapted to the emergence of life and consciousness. This constitutes an argument (but not proof, as it is possible that an infinity of universes exist, each with different characteristics) in favor of the existence of a creator.

In biology, Christian de Duve simultaneously rejects the opinion of those who affirm that natural laws cannot explain the appearance and development of life, as well as the arguments of those who affirm that life and the products of evolution are the result of a totally contingent process. In contrast to a vision that was widespread among numerous biologists (according to whom if evolution began again on a planet identical to Earth, it would produce completely different results), Christian de Duve supports the idea that, where conditions allow it, the laws of nature lead not only to the appearance of life but also to evolved forms of consciousness. Thus, the constraints exerted by these laws of nature on life mean that, even if mutations take place purely by chance, our existence is part of a process and is thus not without meaning.

The third approach is of a methodological nature. It insists (to the surprise of some) on the fact that the approaches of the scientist and the believer are a lot closer than we may think. Indeed, a scientific researcher should, at the outset, have a certain faith—faith in the intelligibility of the world, faith in the fact that an order exists, that the world is not pure chaos but at least partly comprehensible. The (good) researcher should be humble when he or she studies the world and be ready to change his or her mind in the search for truth. If facts contradict his theories, he should make these known to his colleagues, even if that may hinder his career.

Yet what we see is that all these values—the search for truth, humility, faith in the intelligibility of the world—are also those of the (good) seeker of God, or the divine for nonmonotheists. The strong parallelism between these two approaches is what Charles Townes, Jean Kovalevsky, and Thierry Magnin analyze here. They make a particularly strong case for the inexistence of what Jean Bricmont calls "irreducible antagonism between science and religion."[7]

My fourth and last point concerns the basis of the subject itself: does science really tell us anything about meaning (the meaning of our existence, the meaning of the existence of the universe)? It will not escape the attentive reader's attention that a fundamental contradiction exists at the very heart of this work between two very different schools of thought, about the way in which science and meaning interact.

Paul Davies represents the first school, following in the tradition of Einstein. The universe has meaning because we are able to understand it, because there is a link between our spirit and the structure of the universe (or the spirit of its creator for those who believe in God). The second, represented by Bernard d'Espagnat and Thierry Magnin, would say that the universe has meaning because we *cannot* understand it, cannot "unveil" it completely. Because science itself shows us that there is more beyond that which science can teach us. Certainly, it is positive that the "new paradigm" rests on diverse and varied approaches. We should nonetheless be aware of the need for a certain coherence.

Let us push these two ways of reasoning to their limit. Let us first suppose that the world is perfectly comprehensible, that a "theory of everything" explains the reason why the universe has exactly the characteristics that it has and why life, and even man, has appeared. Materialists would hasten to conclude that we have no need for God or a creating principle to explain the world. Everything can be perfectly explained by itself. Believers would reply that this is the proof that nature corresponds with a project that has been carefully programmed down to the smallest details.

Now let us put ourselves in the opposite situation, where science demonstrates the arbitrary character of the laws of nature, with certain values more than others appearing unpredictably out of the ambient chaos. This is proof that the universe is not based on any project, materialists would conclude. But for such an order to appear from such chaos, believers would hasten to reply, it is necessary that God, without violating the laws of nature, has oriented the evolution of the universe.

From these debates we may conclude that "everything is the same," that the links between science and meaning are of a completely subjective nature since, in all cases, all the opinions are as valid as each other. Thus the discipline that studies the evolution of relationships between science and meaning simply does not exist and there is nothing new under the Sun. However, I do not believe this to be the case. It seems, moreover, that Michael Heller's text shows us the way to get out of this deadlock. In the last pages of his text, Heller attempts to show (certainly in a partial way that deserves development, considering the importance of the subject) that we can consider the two approaches *at the same time.*

Our understanding of the world is sufficiently extraordinary for us to be able to see in it the sign of a connection between the human spirit and the spirit of the conceiver of the universe. But the idea that another level of reality also exists, situated outside of time, space, energy, and matter, will reinforce (and not contradict) the idea that the universe has meaning. We are in a situation where, at the same time, we can understand the part of the universe that is accessible to us and where other dimensions exist beyond our understanding that will provide a space for that which is at the origin of the "project" of which our level of reality is the realization.

That is how we can conceive, beyond superficial oppositions, the existence of a global coherence behind the analysis of the scientific evolutions and the philosophical comments that are presented here. It is up to the reader to judge whether this constitutes something "new under the Sun" and whether all of this is a sign of rupture in an era when science and questions of meaning have appeared to be on two different planets. In any case, it seems we can rejoice in the fact that this book represents a valuable testimony to the evolution of ideas concerning the universe and the place that humanity occupies in it at the beginning of the third millennium.

NOTES

1. Steven Weinberg, *The First Three Minutes* (New York: Basic Books, 1977).

2. Jacques Monod, *Chance and Necessity* (Paris: Le Seuil, 1970).

3. Francis Crick, *The Astonishing Hypothesis* (New York: Macmillan, 1994).

4. S. A. Eddington, *The Philosophy of Physical Science* (Cambridge: Cambridge University Press, 1939).

5. E. Schrodinger, *Science & Humanism: Physics in Our Time* (Cambridge: Cambridge University Press, 1951).

6. Ilya Prigogine and Isabelle Stengers, *La Nouvelle Alliance* (Paris: Gallimard, 1979).

7. Jean Bricmont, "Science et religion : l'irréductible antagonisme" in Jean Dubessy, ed., *Intrusions spiritualists et impostures intellectuelles en sciences* (Paris: Editions Syllepse, 2001).

PART I

Science with Philosophy

Revisiting the Paths to Meaning

BERNARD D'ESPAGNAT

Traditional conceptions about existence quite naturally endowed life with meaning. We know, however, that the foundations upon which such conceptions were developed have been—and continue to be—greatly undermined by a certain form of enlightened criticism based on elements of knowledge that large sections of the population in our developed societies think they possess. We shall explain how the findings of contemporary physics are, in turn, undermining the very basis of this criticism and, thereby, reopening perspectives (some of them old and familiar, others quite new) on the pathway to meaning.

Introduction

We all accept that the verb *to understand* has two meanings. When I claim to understand the proof of a theorem, it is not the same as when I grant (as I do!) that I should understand my fellow man. In the first instance, I am referring to an intellectual process, in the second, to an emotional state. The word meaning is subject to the same polysemy even if the intended sense is usually revealed by the context in which it is used. Thus, for example, when considering broad questions of meaning (meaning of action, life, etc.) the "meaning" in this context clearly has little to do with the understanding of a theorem. It follows that when pondering such questions as the meaning of existence in general—as we propose to do in this book—it is to the second definition of the word that we should turn.

When we reflect on questions such as these, we soon notice that a kind of instinctive force has always driven human beings to transpose our understanding of our fellow man to the totality of what is. What I mean is that human beings have steadily strived to grasp the Great All—Being—more or less in this way. From the awakening of Buddha to the immense philosophical resonance of the biblical phrase "I am the One who is," we can retrace the thread of this intuition, which has largely consisted in extrapolating the innate impression of meaningfulness experienced in our daily lives (parents' concern for their children has an immediate and undeniable meaning) from our fellow men—that is to say, from the "biological" or the "relative"—to the absolute.

Today, however, things are quite different and not so straightforward. In the past, such extrapolation or transposition seemed natural as man perceived himself to be immersed within a vast and wondrous world the awesomeness of which stirred his feelings. Woods, the dark of night, clouds, and oceans were the basic elements of his lived experience. They inspired in him a sense of mystery and the infinite that led him, quite naturally, to derive an understanding of the world in its entirety in the second, emotional sense of the word. For the most part, our contemporaries live in towns or cities that are filled with artifacts. We are surrounded by, and constantly come into contact with, tools and mechanisms—clockwork is the paradigmatic example—fashioned by other men and that, as a result, are understandable and analyzable with the help of relatively straightforward concepts. Consequently, our understanding of them is in the first (the intellectual, descriptive) sense of the word. Like our predecessors, we are instinctively driven to transpose our lived, daily experience to the universe as a whole. But, in light of what we have just seen, the very nature of this experience incites us to adopt a totally mechanistic conception of what is, ruling out as absurd any notion of anything fundamentally nontrivial and thereby blocking any hope of understanding the concept of Being according to the second meaning of "to understand."

It is true that for several centuries this blocking effect was countered by dualism. René Descartes, the father of philosophical mechanicism, was fundamentally dualist. He attributed mechanicism merely to matter, not to mind, so that the latter could still be seen as the citadel of meaning. But later scientists started to turn their interests to living bodies, neuronal systems, etc. Quite naturally (and most successfully) they carried over their mechanicist conceptions to these new areas of investigation. When a conceptual model proves to have effective and far-reaching applications, we are normally tempted to use it as a yardstick for the measure of truth itself, as a description of what is. Mechanicism, raised to the status of ontology, therefore gave many scientists the idea that it would eventually be possible to understand everything. Moreover, this would be achieved within the framework of a model that would be free of any serious conceptual problems. This reductionist view naturally led to the undermining of more subtle philosophies (indeed, why develop complex approaches, foreign to our normal experience, when everything may be smoothly explained by directly extrapolating familiar concepts?). But what possible significance can the words "value" or "meaning of existence" have if the entire universe, ourselves included, is merely a collection of machines and can be conceived of as an assemblage of tiny grains or specks linked together by forces? Concerning the question of meaning, generalized mechanicism caused a blockage. Whether implicitly or explicitly, it constituted the basis of the claims that ensued about the loss of meaning and were formulated by many prominent thinkers.

Falsity of Mechanicism

In light of this, it seems to me that anyone who should set out to "recover" meaning in a nonpuerile, nonsuperficial way (that is, without succumbing to wishful thinking) would be well-advised to start by inquiring into the veracity of the mechanicist conception. I do not mean its veracity as a useful model (which is undeniable) or a frequently indispensable tool for action, etc., but rather its "ontological" veracity. In other words, such a person should start by inquiring into the degree of scientific plausibility of the claim that mechanism constitutes the necessary basis of any rationally acceptable conception of what exists—a claim that implies that the possibility of there being anything deeper is ruled out.

To this end, and if this person is truly intent on carrying out his or her quest in a serious, intellectually honest manner, he or she should, of course, first try to get information from physics, since physics (the "science of matter," as it is often called) is the one science upon which any conceivable mechanism must rest. In so doing, this person will find out a nontrivial fact. She will discover that, contrary to superficial appearances, contrary even to what we all believe we learned at school (or in our professional lives), the above-mentioned plausibility is, in fact, equal to zero. Strange as it may seem, the tacit but unavoidable implications of the discoveries of contemporary physics imply that mechanism is only an appearance. As a simple way of representing the apparent behavior of fairly "large" things (even including objects as tiny as the complex molecules that molecular biology deals with), it is wonderfully effective. But we all know that the effectiveness of a model is not a proof of its veracity. Consider, for example, the geocentric model of the world. It was a most useful representation, which led to accurate predictions regarding a significant number of phenomena. Yet, nowadays nobody would think of using it as a valid description of the truth. The same is true of mechanism. As a model it is excellent. However, as an ontology (that is, as a description of the ultimate nature of things), it is, to repeat, false. It is no longer defensible in any form whatsoever. As a result, concerning the question of "Reality"—or Being—subtler philosophical speculations are in order. A priori, it is conceivable that, among them, there should exist some valid ones that will reinstate meaning.

Of course, I am not claiming here that the failure of mechanism de facto restores meaning. We are still at the start of our quest, and the question of meaning is all the more tricky as the very notion of meaning differs from one person to the next. At the stage we are at, only one point is firmly established. It is that, today, all the scientific theories that propose to go beyond the merely utilitarian level have to make use of concepts and modes of thinking that cannot be modeled using notions borrowed from day-to-day life. Consequently, to progress on solid grounds we will have to put aside the question at hand temporarily. Before we go on investigating it, we must take the time to get acquainted with certain essential traits of the new theories in question. It is only after this detour (which we shall limit to the essential) that we shall be able to pursue our quest for the "meaning of meaning."

The Einsteinian View and the Extent to Which the Quantum Approach Made It Obsolete

At first, the "new" scientific theories just alluded to (the theories that were developed during the first half of the twentieth century) were expected, like all earlier ones, to yield a genuine description of the world. In philosophical terms they were expected to provide us with what is generally referred to as an ontology. However, because of new theoretical and experimental developments, it soon turned out that this aim could not be reached merely with the basic notions of mechanism. Hence, a number of fundamental concepts, definable only with the help of mathematics, had to be introduced. In relativity theory, for example, the basic concepts are not those of matter, atoms, particles, etc. They are the various elements of a mathematical structure that calls upon notions such as the relativity of time, the curvature of space, etc. Now, a mathematical structure is not something that, on the face of it, looks radically different from thought. As some distinguished theoretical physicists say, it is a *logos*. And what the theory of relativity suggests, at least to some, is that the *logos* is in reality a world of its own, entirely constituted of subtle symmetries and harmonies.

We shall return to this beautiful and profound vision later. Here we should simply note that it does not represent or, rather, no longer represents, the "last stage" in the conception of knowledge that seems to be progressively emerging from our science. Einstein's relativity was developed in the context of what we now refer to as classical physics.

But, as everyone knows, concerning the study of molecules, atoms, particles, etc. (and also macroscopic objects when investigations are pushed far enough), classical physics turned out to be flawed and had to be replaced with quantum physics, the axioms of which are altogether different. When we study the latter in depth and consider in detail how these axioms are applied in the analysis of observed physical effects, we note that something differentiates them in a fundamental way from those of classical physics, as well as from those of the other sciences. This "something" is the fact that they are essentially constructed as rules for the prediction of observations. Consequently, they are not really descriptive, in the sense of a description of an external reality conceived of as being totally independent of the structures of our mind. It follows that if quantum physics is unsurpassable and universal, as many clues seem to suggest, then science gives us genuine access, not to reality in the ontological sense of the word, but merely to the links between phenomena. In relation to our ordinarily experienced "empirical reality" (i.e., the set of the phenomena), reality as such (i.e., ontological Reality) cannot, therefore, be thought of otherwise than as being some sort of a hyper-reality, unknowable as it really is. In these conditions, the logos that the mathematician finds himself exploring as it exists in itself is clearly neither identifiable to this hyper-reality nor to empirical reality.

An "Allowed Space" of Conceivable Conjectures

Prudence is needed, however. From the fact that Being cannot be unveiled through scientific means, it does not follow that science has nothing to say about it. The truth, of course, is that first and foremost, science informs us reliably on the nature of empirical reality, that is, what concerns us in practice. But, beyond this, the truth is also that even though Nature, in the most fundamental sense of the word, refuses to let us know what she really is, yet, if we probe her insistently enough, she eventually concedes to let us know a little about what she is not. In less figurative terms I would posit that, in the field of possible conjectures about Being, physics defines a kind of "allowed space." The conjectures that lie outside this allowed space are not necessarily absurd but (at least in my view as a physicist) they are highly artificial and implausible. Whereas, conversely, those that lie within the space in question seem to me to be acceptable and even, in some cases, appealing.

Due to limited space, let me say only a few words concerning this line of demarcation. The key point is this: from the fact that our understanding does not reveal to us Reality as it really is, it follows, to repeat, that our physical laws only apply to phenomena, not to this ultimate Reality. Hence, in particular, there is no reason that the temporal evolution described by these laws should be the temporal evolution of Reality. Indeed, it is possible, nay, even likely, that time is not a reality per se, that it is nothing more, in the end, than a human representation. And this also holds true concerning the "cosmic time" of astrophysicists. In other words, there are good reasons to believe that Being, ultimate Reality, is eternal in the etymological sense of the term, that is, prior to time.

This important indication brought to us by contemporary physics should, it seems to me, require us to denounce the confusion that is frequently made between the concepts of *eternity* and *immortality*. The word "immortality" evidently refers to a particular type of evolution in time (just as "being at rest" is a particular type of motion). If time is nothing but a human representation, then so too is immortality—a conclusion that runs against everything that the terms "immortality" and its "harmonics" (afterlife, etc.) are normally meant to signify. It is the notion of "eternity," in the etymological sense of the word, that makes it possible for us to escape this conceptual vicious circle, since it implies that time itself is relative and that what is eternal is "what really is." If this view is correct, then what the tympanums of our cathedrals really represent is the emergence of the saints from the constraints of time and phenomena and their access to Being itself—this being the case even if the conscious purpose of their authors was to describe events in time.

Consequently, only present in this allowed space of mine are the ontological conjectures that conceive of Being as prior to time. And, correlatively, only present in it are the conceptions according to which empirical reality—the set of the phenomena, the reality we are immersed in—is not ultimate reality. In other words, conceptions, according to which true "existence" is the privilege of a timeless suprareality, are quite distinct from empirical reality, the latter being, as one might say, just a trace of the former.

Back to the Question of Meaning

The rapid survey we have just made of the data—or rather, let us say, the "indications" yielded by present-day physics—will now be made use of in a quest for meaning that, hopefully, will be self-consistent and circularity free. For the sake of clarity, I shall split it into two relatively independent parts. In the first, we shall start from a conception that could be termed "the received religious approach to the question." We shall consider the arguments for and against it and try to find out if and to what extent its presuppositions are compatible with at least one of the ontological conjectures present within the above-mentioned "allowed space." In the second one, we shall examine whether or not, in the light of present-day knowledge, it is possible to build up a conception of meaning that is free from any traditional preconceptions and, nevertheless, inspiring.

The Religious Starting Point

In the traditional religious way of thinking, the apprehension of a "meaning" (of life, etc.) is linked to the notion of salvation and thereby to that of an afterlife. This, we may say, is its first characteristic. The second characteristic calls upon the notion of a God who expects something of us, that is, One endowed with an attribute akin to will.

Taken literally, the afterlife notion is, as we have seen, problematic in my eyes, since it implies that time has an absolute existence. Let us keep in mind, however, that this difficulty may be removed in a satisfactory manner. This merely requires substituting the idea of an afterlife taking place within the time our clocks measure (a most naive view anyhow!) with the idea of a genuinely eternal, that is, time-free, existence. As for the notion "God," in the light of contemporary physics it no longer raises any insurmountable problem so long as (in the lines of Descartes and many others) we simply identify him with Being. (It no longer raises a problem since contemporary physics indicates, as we have seen, that Reality-per-se, alias Being, is not to be confused with empirical reality, is prior to time, and so on.) On the other hand, it must be granted that to a person pondering these problems from the perspective of a physicist, it may, at first sight, seem intolerably artificial to attribute to Reality-per-se anything resembling will, intention, etc.; in short, to identify it with God in the usual sense of the word (or even, less specifically, to identify it with "the Godhead"; the distinction between "God" and "the Godhead" is necessary, it seems to me, since the notion of design, or intention, does not necessarily imply that "that" in which the intention resides is, in any way, a "person").

On closer inspection, however, it seems to me that the attribution in question, while speculative, is less unacceptable than might at first appear to be the case. Indeed, in view of some features of contemporary science, I consider it likely that, far from being a mere emanation of empirical reality, consciousness emerges—either prior to the latter or simultaneously with it as the other side of one and the same medal—from Being itself (it being understood that the expressions "prior to" and "simultaneously with" refer to a conceptual as opposed to a temporal ordering).[1] In this case, if one of these emergent entities—empirical reality—preserves (as seems to be the case) some traces of Being, it is natural to conjecture that the same holds true concerning the other one, that is, consciousness. Hence, the idea becomes plausible according to which Being itself possesses attributes, of which some features of our minds (including the ability to have intentions) yield an image that is not altogether misleading (even though the image in question is undoubtedly most imperfect and deformed).

I am not the only physicist to hold such ideas. In the later part of his career, David Bohm, as we know, conjectured the existence of a common foundation—profound and hidden—to matter and mind, coinciding with neither one nor the other. And in one of his interviews with Renée Weber,[2] in response to a question, he went so far as to say that since the foundation in question is at the same time that of matter and mind, it is presumably endowed with an awareness of some kind. If we grant this, then it is natural to consider that the said foundation has a (timeless) attribute of which our capacity to desire, and our quest of the good, are a dim reflection. The idea that it may require—or simply expect—this or that from us then enters the realm of the conceivable.

An Approach Freer from Cultural Preconceptions

As already noted, physicists and scientists in general are seldom inclined to endow what "basically exists," (Being) with any attribute such as will, love, or intention that has anything to do with the mind. This has not always been the case. It seems that the idea of identifying God (the God of Christianity) with Being posed no real difficulty to Descartes. But still, when present-day scientists think of what basically exists, they spontaneously think of the universe, which, apparently, neither thinks nor feels anything. With respect to the question of meaning, this raises a serious problem since—as many would ask—however immense a pack of galaxies—which neither feel nor wish—may be, what motivations, what warmness can it raise in us? Didn't Pascal point out its fundamental inferiority with respect to the "thinking reed" that man is?

With many of our contemporaries—scientists and nonscientists—this way of looking at things constitutes a serious stumbling block on the pathway toward meaning. It is worth noting, however, that this is less often the case among theoretical physicists. Indeed, the latter know that if "what really is" is, in some way, representable, it can only be via the channel of mathematics. We have already noted that mathematical structures come close to thought. Naturally, the importance of their role incites many theoretical physicists to adopt, explicitly or not, a Platonic view of what *is*: to entertain a conception according to which what exists in the first instance is the *logos* we referred to. According to them, it is from this world of pure mathematics that we should derive our understanding of the physical world. But then, the *logos* is made up of subtle symmetries and harmonies that make it a receptacle of beauty. Clearly, the scientist who glimpses this world and its beauty is not, in relation to it, in a situation differing very much from that of the scientists of the past, who first glimpsed the mysteries of nature. Like them, he or she naturally considers that it makes sense to embark with a sense of wonder on an exploration of it. Einstein, for one, had an intense experience of such a state of mind and strived to convey it while at the same time generalizing it. Man, he wrote, experiences the inanity of desires and human objectives and the sublime and wonderful character of the order revealed in nature and the world of thought. He views his individual existence as a sort of prison and wants to live the totality of what is something endowed with both unity and meaning.[3]

Such an approach to meaning is not available to everybody. And it remains true that we find it difficult to build up for ourselves, on the basis of a conception of the "ground of things" stripped of any intentionality whatsoever, a notion of meaning that should be an inspiring one. But still, if we think again of these men of yore to whom contemplation of the visible universe quite naturally revealed meaning, we may find out that, after all, for them as for Einstein, the driving impetus was not, in the final analysis, the idea that the universe in question is endowed with will or design. To them, it seems to me, such a notion, even when loudly put forward by their thinkers, sounded, in truth, secondary—just a seemingly plausible interpretation of the immeasurable majesty of the world. I think that, today, the notion of a Being-in-itself—of a hyper-Real—lying at the source of all existence and vastly escaping our cognitive capacities, can, without demanding a shift of meaning, inspire in us the same feelings.

To conclude, I consider that, in spite of an undeniable obscurity inherent in the theme, we may discern at least two conceptual pathways making it possible to respond positively to the question of meaning. One of them—the one we considered last—should satisfy people who judge that scientific strictness is hardly compatible with mere conjectures, even in an area in which it is generally acknowledged that science merely sets safeguards. The other one, closer to the spiritual tradition, is better adapted to the mentality of those who do not experience—or experience less—this type of reticence, but who, on the other hand, have a more restrictive vision of what can truly open the road to meaning. Selecting one or the other of these two options is, I think, a matter of personal choice.

NOTES

1. See e.g. Bernard d'Espagnat, "On consciousness and the Wigner's friend problem," *Foundations of Physics* 35, no. 12 (December 2005): 1943. See also Bernard d'Espagnat, *Traité de physique et de philosophie* (Paris: Fayard, 2002); English translation, *On Physics and Philosophy* (Princeton: Princeton University Press, 2006).

2. In Renée Weber, *Dialogues with Scientists and Sages* (London: Routledge and Kegan Paul, 1986), 95.

3. Albert Einstein, *Mein Weltbild* (Amsterdam: Querido Verlag, 1934).

Glimpsing the Mind of God

PAUL DAVIES

We live, it is said, in the Scientific Age. Members of the public, and even most scientists, take science for granted. They expect it to work. But why is science so successful in describing our world, and how is it that human beings have evolved the capability of understanding the deep principles on which the universe runs?

Of course, science didn't spring ready-made into the minds of its founders like Galileo, Descartes, and Newton. They were strongly influenced by two longstanding traditions that pervaded European thought: Greek philosophy and the Judaic worldview. In most ancient cultures, people were aware that the universe is not completely chaotic and capricious; there is a definite order in nature. The Greeks believed that this order could be understood, at least in part, by the application of human reasoning. They maintained that physical existence was not absurd but rational and logical and, therefore, in principle intelligible to us. They discovered that some physical processes had a hidden mathematical basis, and they sought to build a model of reality based on arithmetical and geometrical principles.

The second great tradition was the Judaic worldview, according to which the universe was created by God at some definite moment in the past and ordered according to a fixed set of laws. The Jews taught that the universe unfolds in a unidirectional sequence—what we now call linear time—according to a definite historical process: creation, evolution, and dissolution. This notion of linear time (in which the story of the universe has a beginning, a middle, and an end) stands in marked contrast to the concept of cosmic cyclicity, the pervading mythology of almost all ancient cultures. Cyclic time—the myth of the eternal return—springs from mankind's close association with the cycles and rhythms of nature, and remains a key component in the belief systems of many cultures today. It also lurks beneath the surface of the Western mind, erupting occasionally to infuse our art, our folklore, and our literature.

A world freely created by God, and ordered in a particular, felicitous way at the origin of a linear time, constitutes a powerful set of beliefs, and was taken up by both Christianity and Islam. An essential element of this belief system is that the universe does not *have* to be as it is; it could have been otherwise. Einstein once said the thing that most interested him is whether God had any choice in the form of his creation. According to the Judaeo-Islamic-Christian tradition, the answer is yes.

Although not conventionally religious, Einstein often spoke of God and expressed a sentiment shared, I believe, by many scientists, including professed atheists. It is a sentiment best described as a reverence for nature and a deep fascination for the natural order of the cosmos. If the universe did not have to be as it is, of necessity—if, to paraphrase Einstein, God did have a choice—then the fact that nature is so fruitful, that the universe is so full of richness, diversity, and novelty, is profoundly significant. The fact that it is also *intelligible* to at least one species on one planet is also profoundly significant.

Some scientists have tried to argue that if only we knew enough about the laws of physics, if we were to discover a final theory that united all the fundamental forces and particles of nature into a single mathematical scheme, then we would find that this superlaw, or theory of everything, would describe a unique, logically consistent world. In other words, the nature of the physical world would be entirely a consequence of logical and mathematical necessity; there would be no choice about it. I think this is demonstrably wrong. There is not a shred of evidence that the universe is logically necessary. Indeed, as a theoretical physicist,

I find it rather easy to imagine alternative universes that are logically consistent and, therefore, equal contenders for reality.

It was from the intellectual ferment brought about by the merging of Greek philosophy with Judaeo-Islamic-Christian thought that modern science emerged, with its unidirectional linear time, its insistence on nature's rationality, and its emphasis on mathematical principles. All of the early scientists, such as Newton, were religious in one way or another. They saw their science as a means of uncovering traces of God's handiwork in the universe. What we now call the laws of physics were regarded as God's abstract creation, thoughts, so to speak, in the mind of God. In doing science, they supposed, one might be able to glimpse the mind of God. What an exhilarating and audacious claim!

In the ensuing three hundred years, the theological dimension of science has faded. People take it for granted that the physical world is both ordered and intelligible. The underlying order in nature—the laws of physics—are simply taken by most scientists as given, as brute facts. The lawlike order in nature that is at least in part comprehensible to us is accepted as an act of sheer faith.

It has become fashionable in some circles to argue that science is ultimately a sham, that we scientists read order into nature, not out of nature, so that the laws of physics are our laws, not nature's. I believe this is nonsense. You'd be hard pressed to convince a physicist that Newton's inverse square law of gravitation is a purely cultural concoction. The laws of physics, I submit, *really exist* in the world out there, and the job of the scientist is to uncover them, not to invent them. True, at any given time the laws you find in the textbooks are tentative and approximate, but they mirror, albeit imperfectly, a really existing order in the physical world. Of course, many scientists don't recognize that in accepting the reality of an order in nature—the existence of laws "out there"—they are adopting a theological worldview.

Let us accept, then, that nature really is ordered in a mathematical way—that "the book of nature," to quote Galileo, "is written in mathematical language." Even so, it is easy to imagine an ordered universe that nevertheless remains utterly beyond human comprehension due to its complexity and subtlety. For me, the magic of science is that we *can* understand at least part of nature—perhaps, in principle, all of it—using the scientific method of enquiry. How utterly astonishing that we human beings can do this! Why should the rules on which the universe runs be accessible to the human intellect?

The mystery is all the greater when one takes into account the cryptic character of the laws of nature. When Newton saw the apple fall, he saw a falling apple. He didn't see a set of differential equations that link the motion of the apple to the motion of the Moon. The mathematical laws that underlie physical phenomena are not apparent to us through direct observation; they have to be painstakingly extracted from nature using arcane procedures of laboratory experiment and mathematical theory. The laws of nature are hidden from us and are revealed only after much labor. The late Heinz Pagels described this by saying that the laws of nature are written in a sort of cosmic code, and that the job of the scientist is to crack the code and reveal the message—nature's message, God's message, take your choice, but not *our* message.[1] The extraordinary thing is that human beings have evolved such a fantastic code-breaking talent. This is the wonder and the magnificence of science: we can use it to decode nature and discover the secret laws that make the universe tick!

Many people want to find God in the *creation* of the universe, in the big bang that started it all off. They imagine a superbeing who deliberates for all eternity, then presses a metaphysical button and produces a huge explosion. I believe this image is entirely misconceived. Einstein showed us that space and time are *part of* the physical universe, not a preexisting arena in which the universe happens. In the simplest model of the big bang theory, the origin of the universe represents the coming-into-being not just of matter and energy, but of space and time as well. Time itself began with the big bang. If this sounds baffling, it is by no means new. Already in the fifth century Saint Augustine proclaimed that "the world was made with time, not in time." According to James Hartle and Stephen Hawking, this coming-into-being of the universe need not be a supernatural process but could occur entirely naturally in accordance with the laws of quantum physics, which permit the occurrence of genuinely spontaneous events.[2]

The origin of the universe, however, is hardly the end of the story. The evidence suggests that in its primordial phase the universe was in a highly simple, almost featureless state—perhaps a uniform soup of sub-atomic particles or even just expanding empty space. All the richness and diversity of matter and energy we observe today has emerged since the beginning in a long and complicated sequence of self-organizing physical processes. What an incredible thing these laws of physics are! Not only do they permit a universe to originate spontaneously, but they encourage it to self-organize and self-complexify to the point where conscious beings emerge who can look back on the great cosmic drama and reflect on what it all means.

Now you may think I have written God entirely out of the picture. Who needs a God when the laws of physics can do such a splendid job? But we are bound to return to that burning question: Where do the laws of physics come from? And why *those* laws rather than some other set? Most especially, why a set of laws that drives the searing, featureless gases coughed out of the big bang toward life and consciousness and intelligence and cultural activities such as religion, art, mathematics, and science? If there is a meaning or purpose to existence, as I believe there is, we are wrong to dwell too much on the originating event. The big bang is sometimes referred to as "the creation," but in truth, nature has never *ceased* to be creative. This ongoing creativity, which manifests itself in the spontaneous emergence of novelty, complexity, and organization of physical systems, is permitted through or guided by the underlying mathematical laws that scientists are so busy discovering.

Now the laws of which I speak have the status of timeless eternal truths, in contrast to the physical states of the universe that change with time and bring forth the genuinely new. So we here confront in physics a reemergence of the oldest of all philosophical and theological debates, the paradoxical conjunction of the eternal and the temporal. Early Christian thinkers wrestled with the problem of time. Is God within the stream of time or outside of it? How can a truly timeless God relate in any way to temporal beings such as ourselves? But how can a God who relates to a changing universe be considered eternal and unchangingly perfect? Well, physics has its own variations on this theme. In our century, Einstein showed us that time is not simply "there" as a universal and absolute backdrop to existence; it is intimately interwoven with space and matter. As I have mentioned, time is revealed to be an integral part of the physical universe; indeed, it can be warped by motion and gravitation. Clearly, something that can be changed in this manner is not absolute, but a contingent part of the physical world.

In my own field of research, called quantum gravity, a lot of attention has been devoted to understanding how time itself could have come into existence in the big bang. We know that matter can be created by quantum processes. There is now a general acceptance among physicists and cosmologists that space-time can also originate in a quantum process. According to the latest thinking, time might not be a primitive concept at all but something that has "congealed" from the fuzzy quantum ferment of the big bang—a relic, so to speak, of a particular state that froze out of the fiery cosmic birth.

If it is the case that time is a contingent property of the physical world rather than a necessary consequence of existence, then any attempt to trace the ultimate purpose or design of nature to a *temporal* Being or Principle seems doomed to failure. While I do not wish to claim that physics has solved the riddle of time (far from it), I do believe that our advancing scientific understanding of time has illuminated the ancient theological debate in important ways. I cite this topic as just one example of the lively dialogue that is continuing between science and theology.

So where is God in this story? Not especially in the big bang that starts the universe off, nor meddling fitfully in the physical processes that generate life and consciousness. I would prefer the idea that nature takes care of itself. The idea of a God who is just another force or agency at work in nature, moving atoms here and there in competition with physical forces, is profoundly uninspiring. To me, the true miracle of nature is to be found in the ingenious and unswerving lawfulness of the cosmos, a lawfulness that permits complex order to emerge from chaos, life to emerge from inanimate matter, and consciousness to emerge from life, without the need for the occasional supernatural prod; a lawfulness that produces beings who not only ask great questions of existence but who, through science and other methods of enquiry, are even beginning to find answers.

You might be tempted to suppose that any old rag-bag of laws would produce a complex universe of some sort, with attendant inhabitants convinced of their own specialness; not so. It turns out that randomly selected laws lead almost inevitably to either unrelieved chaos or boring and uneventful simplicity. Our own universe is poised exquisitely between these unpalatable alternatives, offering a potent mix of freedom and discipline, a sort of restrained creativity. The laws do not tie down physical systems so rigidly that they can accomplish little, nor are they a recipe for cosmic anarchy. Instead, they encourage matter and energy to develop along pathways of evolution that lead to novel variety, what Freeman Dyson has called the principle of maximum diversity—that in some sense we live in the most interesting possible universe.[3]

Scientists have recently identified a regime dubbed "the edge of chaos," a description that certainly characterizes living organisms, where innovation and novelty combine with coherence and cooperation. The edge of chaos seems to imply the sort of lawful freedom I have just described. Mathematical studies suggest that to engineer such a state of affairs requires laws of a very special form. If we could twiddle a knob and change the existing laws, even very slightly, the chances are that the universe as we know it would fall apart, descending into chaos. Certainly, the existence of life as we know it, and even of less elaborate systems such as stable stars, would be threatened by just the tiniest change in the strengths of the fundamental forces, for example. The laws that characterize our actual universe, as opposed to an infinite number of alternative possible universes, seem almost contrived—fine-tuned, some commentators have claimed—so that life and consciousness may emerge. To quote Dyson again, it is almost as if "the universe knew we were coming."[4] I can't prove to you that that is design but, whatever it is, it is certainly very clever!

Now some of my colleagues embrace the same scientific facts as I but deny any deeper significance. They shrug aside the breath-taking ingenuity of the laws of physics and the extraordinary felicity of nature as a package of marvels that just happens to be. Alternatively, they embrace the so-called multiverse theory, explained by Hubert Reeves. According to this point of view, what we call "the universe" is but an infinitesimal component of a much larger system.[5] There are other regions of space, or entire universes existing in parallel to ours, in which the laws are different. Perhaps all possible laws are manifested in a universe somewhere. But only in a tiny fraction of universes will the laws be bio-friendly, permitting life to emerge with beings such as ourselves who can study their world and marvel at how propitiously the laws have been arranged. However, they would be wrong to read any significance into this observation, for such observers are merely winners in a vast cosmic lottery. The other, less hospitable universes, will go unseen.

The multiverse theory seems to have become the explanation of choice among scientists for the remarkable bio-friendliness of the observed universe, but I have problems with it. First, it is necessary to assume that all universes have laws of some sort. We still have to explain "lawfulness." Second, if the multiverse theory is right, we should live in the *least* contrived and bio-friendly universe consistent with the emergence of intelligent life, because there will be many more universes where things are just contrived enough than universes that are even more contrived. There is no evidence that our universe "just crosses the wire" when it comes to ingenious bio-friendly features. Third, the ontological status of the multiverse theory is, I submit, isomorphic to classical theism. Both theories invoke an infinitely complex, unseen agency to explain the universe we do see; the multiverse appeals to hidden universes, theism to a hidden deity. There is a branch of mathematics, known as algorithmic information theory, that can be used to quantify the complexity of theories according to the information content hidden in their assumptions. I conjecture that, defined in the precise language of algorithmic information theory, the multiverse and classical theism would turn out to be both equally—and infinitely—complex. One might say that the multiverse theory is merely theism dressed up in scientific jargon. My personal hope is that there is a "third way" in which the ingenious, bio-friendly lawfulness of the universe will be explained without appeal to infinitely complex unseen agencies.

If we reject the multiverse explanation, then the laws of the universe point forcefully to a deeper underlying meaning to existence. Some call it purpose, some design. These loaded words, which derive from human categories, capture only imperfectly what it is that the universe is *about*. But that it is about something, I have absolutely no doubt.

Where do we human beings fit into this great cosmic scheme? Can we gaze out into the cosmos, as did our remote ancestors, and declare, "God made all this for us!" Well, I think not. Are we then but an accident of nature, the freakish outcome of blind and purposeless forces, an incidental by-product of a mindless, mechanistic universe? I reject that, too. The emergence of life and consciousness, I maintain, are written into the laws of the universe in a very basic way. True, the actual physical form and general mental makeup of homo sapiens contains many accidental features of no particular significance. If the universe were re-run a second time, there would be no solar system, no Earth, and no people. But the emergence of life and consciousness somewhere and somewhen in the cosmos is, I believe, assured by the underlying laws of nature. The origin of life and consciousness were not interventionist miracles but nor were they stupendously improbable accidents. They were, I believe, part of the natural outworking of the laws of nature and, as such, our existence as conscious enquiring beings springs ultimately from the bedrock of physical existence—those ingenious, felicitous laws. That is why I wrote in my book *The Mind of God*, "We are truly meant to be here."[6] I mean "we" in the sense of conscious beings, not homo sapiens specifically. Thus, although we are not at the centre of the universe, human existence *does* have a powerful wider significance. Whatever the universe as a whole may be about, the scientific evidence suggests that we, in some limited yet ultimately still profound way, are an integral part of its purpose.

How can we test these ideas scientifically? One of the great challenges to science is to understand the nature of consciousness in general and human consciousness in particular. We still do not understand how mind and matter are related nor what process led to the emergence of mind from matter in the first place. Second, if I am right that the universe is fundamentally creative in a pervasive and continuing manner, and that the laws of nature encourage matter and energy to self-organize and self-complexify to the point that life and consciousness emerge naturally, then there will be a universal trend or directionality toward the emergence of greater complexity and diversity. We might then expect life and consciousness to exist throughout the universe. That is why I attach such importance to the search for extraterrestrial organisms, be they bacteria on Mars, or advanced technological communities on the other side of the Galaxy. The search may prove hopeless (the distances and numbers are certainly daunting) but it is a glorious quest. If we *are* alone in the universe, if the Earth is the only life-bearing planet among countless trillions, then the choice is stark. Either we are the product of a unique supernatural event in a universe of profligate overprovision or else an accident of mind-numbing improbability and irrelevance. On the other hand, if life and mind are universal phenomena, if they are written into nature at its deepest level, then the case for an ultimate purpose to existence would be compelling.

I believe that mainstream science, if we are brave enough to embrace it, offers the most reliable path to knowledge about the physical world. I am certainly not saying that scientists are infallible, nor am I suggesting that science should be turned into a latter-day religion. But I do think that if religion is to make real progress, it cannot ignore the scientific culture, nor should it be afraid to do so. For, as I have argued, science reveals just what a marvel the universe is. It is not the plaything of a capricious Deity but a coherent, rational, elegant, and harmonious expression of a deep and purposeful meaning.

NOTES

1. Heinz Pagels, *The Cosmic Code* (New York: Simon & Schuster, 1982).
2. J. Hartle and S. W. Hawking, "Wave function of the universe," *Phys. Rev.* D28, 2960 (1983).

3. Freeman Dyson, "Progress in Religion" (acceptance speech for the Templeton Prize, Washington National Cathedral, Washington, D.C., May 9, 2000).

4. Freeman Dyson, *Disturbing the Universe* (New York: Harper & Row, 1979), 250.

5. A general reference to the multiverse is Bernard Carr, ed., *Universe or Multiverse?* (Cambridge: Cambridge University Press, 2006).

6. Paul Davies, *The Mind of God* (New York: Simon & Schuster, 1992), 232.

3

Mysteries of Life[1]

Is There "Something Else"?

CHRISTIAN DE DUVE

Introduction

Science is based on *naturalism,* the notion that all manifestations in the universe are explainable in terms of the known laws of physics and chemistry. This notion represents the cornerstone of the scientific enterprise. Unless we subscribe to it, we might as well close our laboratories. If we start from the assumption that what we are investigating is not explainable, we rule out scientific research.

Contrary to the view expressed by some scientists, this logical necessity does not imply that naturalism is to be accepted as an a priori philosophical stand, a doctrine or belief. As used in science, it is a postulate, a *working hypothesis,* often qualified as methodological naturalism by philosophers for this reason, which we should be ready to abandon if faced with facts or events that defy every attempt at a naturalistic explanation. But only then should we accept the intervention of "something else," as a last resort after all possibilities of explaining a given phenomenon in naturalistic terms have been exhausted. Should we reach such a point, assuming it can be recognized, we may still have to distinguish between two alternatives. Is the "something else" an unknown law of nature now disclosed by our investigations, as has happened several times in the past? Or is it a truly supernatural agency?

Traditionally, life, with all its wonders and mysteries, has been a favorite ground for the belief in "something else." Largely muted by the spectacular advances of biology in the last century, this position has been brought back into prominence by a small but vocal minority of scientists, whose opinions have been widely relayed in various philosophical and religious circles. In this essay, I wish to examine briefly whether certain biological phenomena indeed exist, as is claimed, that truly defy every attempt at a naturalistic explanation and make it necessary to invoke the intervention of "something else." Has a stage been reached where all scientific avenues have been exhausted? Should such be the case, are we to enlarge our notion of what is natural? Or have we met the authentically supernatural?

In examining these questions, I have assumed that readers are familiar with at least the basic elements of modern biology. Also, references have been strictly limited. For additional information and a more detailed treatment of the topics addressed, readers are referred to earlier works.[2]

The Nature of Life

Research into the mechanisms that support life represents one of the most spectacular successes of the naturalistic approach. I can bear personal witness to this astonishing feat. When I was first exposed to biochemistry, only a number of small biological building blocks, such as sugars, amino acids, purines, pyrimidines, fatty acids, and a few others, had been identified. How these molecules were made by living organisms was largely unknown. Not a single macromolecule arising from their combination had been characterized. Of metabolism only a few central pathways, such as glycolysis and the tricarboxylic acid cycle, had been painstakingly unravelled. Enzymology was still in its infancy. So was bioenergetics, which at that time boiled down to the recent discovery of ATP and some hints of the role of this substance as a universal purveyor of biological energy. As to genetic information transfers, our ignorance was complete. We did not even know the function of DNA, let alone its structure.

We were hardly disheartened by the puniness of these achievements.

On the contrary, we saw them as tremendous triumphs. They opened wide vistas, testified to the fact that life could be approached with the tools of biochemistry, and strongly encouraged us to further research. But the unsolved problems loomed huge on the horizon and their solution appeared remote. I, for my part, never imagined in my wildest dreams—and I don't think any of my contemporaries did—that I would live to see them elucidated. Yet, that is what has happened. We now know the structures of all major classes of macromolecules, and we have the means to isolate and analyze any single such molecule we choose. We understand in detailed fashion most metabolic pathways, including, in particular, all major biosynthetic processes. We also have an intimate understanding of the mechanisms whereby living organisms retrieve energy from the environment and convert it into various forms of work. Most impressive of all, we know how biological information is coded, stored, replicated, and expressed. It is hardly an exaggeration to say that we have come to understand the fundamental mechanisms of life. Many gaps in this knowledge remain to be filled. For all we know, some surprises may still await us—remember the discovery of split genes. But, on the whole, *the basic blueprint of life is known,* to the point that we are now capable of manipulating life knowingly and purposefully in unprecedented fashion.

The lesson of these remarkable accomplishments is that life is *explainable in naturalistic terms.* To be true, new principles have been uncovered that rule the behavior of complex molecules such as proteins and nucleic acids, or govern the properties of multimolecular complexes such as membranes, ribosomes, or multi-enzyme systems. But nothing so far has been revealed that is not explainable by the laws of physics and chemistry. These disciplines have merely been greatly enriched by the new knowledge. There is no justification for the view that living organisms are made of matter "animated" by "something else."

The History of Life

Modern biological knowledge has revealed another capital piece of information: all known living organisms are descendants from a *single ancestral form of life.* Already suspected by the early evolutionists, this view has been further bolstered by the close similarities that have been detected at the cellular and molecular levels among all analyzed living organisms, whatever their apparent diversity. Whether we look at bacteria, protists, plants, fungi, or animals, including humans, we invariably find the basic blueprint mentioned above. There are differences, of course. Otherwise, all organisms would be identical. But the differences are clearly recognized as variations on the same central theme. *Life is one.*

This fact is now incontrovertibly established by the sequence similarities that have been found to exist among RNA or protein molecules that accomplish the same functions in different organisms or among the DNA genes that code for these molecules. The number of examples of this sort now counts in the many hundreds and is continually growing. It is utterly impossible that molecules with such closely similar sequences could have arisen independently in two or more biological lines, unless one assumes a degree of determinism in the development of life that even the most enthusiastic supporter of this view would refuse to consider.

These sequence similarities not only prove the kinship among all the organisms that have been subjected to this kind of analysis; they can even serve to establish phylogenetic relationships on the basis of the hypothesis, subject to many refinements, that sequence differences are all the more numerous the longer the time the owners of the molecules have had to evolve separately, that is, the longer the time that has elapsed since they diverged from their last common ancestor. This method, which is now widely applied, has confirmed and strengthened many of the phylogenies previously derived from the fossil record; and it has, in addition, allowed such reconstructions to be extended to the many lineages that have left no identifiable fossil remains.

It is fair to state that the early hopes raised by this powerful new technology have become somewhat tempered. Underlying assumptions have been attacked as oversimplified. Different competing algorithms are advocated as dealing best with the diversity of genetic changes that have to be taken into account. Most important, horizontal gene trans-fer—that is, the transfer of genes between distinct species, as opposed to their vertical transfer from generation to generation—has been rec-ognized as a major complication when attempting to use molecular data to reconstruct the tree of life, especially its early ramifications. These difficulties, however, affect only the shape of the tree, not its reality. *Bio-logical evolution is an undisputable fact.*

The Origin of Life

Where, when, and especially how did life start? We don't have answers to these questions. But, at least, we are no longer completely in the dark about them. We now know, from unmistakable fossil traces, that elabo-rate forms of bacterial life, with shapes reminiscent of photosynthetic or-ganisms known as cyanobacteria, were present on Earth at least 3.55 bil-lion years ago. More primitive organisms must have existed before that date. Considering that the Earth most likely was physically unable to harbor life during at least the first half-billion years after its birth some 4.55 billion years ago, it appears that our planet started bearing living organisms at the latest 500 million years after it became capable of do-ing so.

It is not known whether the first forms of Earth life arose locally or came from elsewhere. Since there is at present no compelling reason or evidence supporting an extraterrestrial origin, most investigators ac-cept the simpler assumption of a local origin, which has the advantage of allowing the problem to be defined within the framework of the physical-chemical conditions, revealed by geological data, that may have prevailed on Earth at the time life appeared. Note that the old argument that not enough time was available on Earth for the natural development of something as complex as even the most primitive living organism is no longer considered valid. It is generally agreed that if life originated naturally, it can only have done so in a relatively short time, probably to be counted in millennia rather than in millions of years. Whatever the pathways followed, chemical reaction rates must have been appreciable in order for fragile intermediates to reach concentrations sufficient to permit the next step to occur. There may thus have been many opportunities on the prebiotic Earth for life to appear and disappear before it finally took root.

Another key piece of evidence has become available in the last thirty years. It has been learned, from the chemical analysis of comets and meteorites and from the spectral analysis of the radiation coming from other parts of the solar system and from outer space, that organic substances, including amino acids and other potential building blocks of life, are widely present in the cosmos. These compounds are most likely products of spontanous chemical processes, not of biological activity. Organic syntheses thus not only do not require living organisms, they can even proceed without human help, and do so on a large scale. Organic chemistry is wrongly named. It is simply carbon chemistry, which happens to be the most banal and widespread chemistry in the universe, while being extraordinarily rich thanks to the unique associative properties of the carbon atom. It seems reasonable to assume that the products of this cosmic chemistry provided the raw materials for the formation of the first living organisms.

The question is: how? Ever since the Russian pioneer Alexander Oparin first tackled the problem in 1924, and especially after Stanley Miller's historic 1953 experiments, some of the best organic chemists in the world have struggled with this question in the laboratory, adopting as basic premise the naturalistic postulate, the hypothesis that the origin of life can be explained in terms of physics and chemistry.[3] Using the same premise, many bystanders, like myself, have speculated on the matter or proposed models. A number of interesting facts have been uncovered, while a much larger number of suggestive ideas or "worlds"— the RNA world, the pyrophosphate world, the iron-sulfur world, the thioester world are examples—have been bandied about. Tens of books, hundreds of scientific papers, a journal specially created for the purpose, a society devoted exclusively to the topic, regular congresses and symposia, all attest to the vitality of this new scientific discipline.

Opinions are divided on what has been accomplished by all this activity. While much has been learned, it is clear that we are still nowhere near explaining the origin of life. This is hardly surprising, considering the immense complexity of the problem. But must the naturalistic postulate be abandoned for that reason? Have we reached a stage where all attempts at a naturalistic explanation have failed and the involvement of "something else" must be envisaged?

Anybody acquainted with the field is bound to answer this question with an emphatic "No." The surface has hardly been scratched. Pronouncing the origin of life unexplainable in naturalistic terms at the present stage can only be based on an a priori surrender to what the American biochemist Michael Behe, a prominent defender of this thesis, calls "irreducible complexity," which he defines in his *Darwin's Black Box,* as the state of "a single system composed of several well-matched interacting parts that contribute to the basic function, wherein the removal of any one of the parts causes the system to effectively cease functioning."[4] As a simple example of an irreducibly complex system, Behe offers the "humble mousetrap," of which each part can have been made only by somebody who had the whole contraption in mind. So it is, according to Behe, with many complex biochemical systems, such as cilia or the blood-clotting cascade.

This argument from design is not new. It was made two centuries ago by the English theologian William Paley, who used the image of a watch to prove the existence of a divine watchmaker. What is new, as well as surprising, is the use of modern biochemical knowledge in support of the "intelligent design" of life, a discovery that Behe believes "rivals those of Einstein, Lavoisier and Schrödinger, Pasteur and Darwin." This is a strange claim for a "discovery" that, instead of solving a problem, removes it from the realm of scientific inquiry.[5]

Mechanical analogies, be they to watches or mousetraps, are poor images of biochemical complexity. Proteins, the main components of biochemical systems, have none of the rigidity of mechanical parts. Their name, derived from the Greek *prôtos* (first) by Gerardus Johannes Mulder, the Dutch chemist who invented it in 1838, could just as well have been derived from the name of the god Prōteus, famous for his ability to take almost any shape. As is well known, replacing a single amino acid by another may completely alter the properties of a protein. Even without any sequence alteration, the shape of a protein may change simply by contact with a modified template, as evidenced by the agents of diseases such as "mad-cow disease" and its human equivalent, Creutzfeldt-Jakob disease. Called prions, these infectious agents consist of abnormally shaped proteins that, according to their discoverer, the American medical scientist Stanley Prusiner, multiply by conferring their abnormal shape to their normal counterparts in the body.[6] To affirm, as is implicit in Behe's theory, that a protein playing a given role cannot be derived from a molecule that fulfilled a different function in an earlier system is a gratuitous assertion, which flies in the face of evidence. Many examples are indeed known of proteins that have changed function in the course of evolution. Crystalins, the lens proteins, are a case in point.

The *time element* is what is missing in Behe's reasoning. It is no doubt true that present-day biochemical systems exhibit what he calls "irreducible complexity."[7] Let a single element of the blood coagulation system be absent, and a major loss of functional efficiency does indeed occur, as evidenced by hemophilia and other similar disorders. But what the argument ignores is that the system has behind it hundreds of millions of years of evolution, in the course of which its slow progressive assembly most likely took place by a long succession of steps each of which could be explainable in naturalistic terms. The fact that the details of this pathway have not yet been unravelled is hardly proof that it could not have happened.

The molecular history of the proteins themselves also needs to be taken into account. Today's proteins are the products of a very long evolutionary history, during which an enormous amount of innovative diversification and adaptation has taken place. Admittedly, even their remote ancestors must already have been of considerable complexity to support the kind of bacterial life revealed by early microfossils. Had those ancestral proteins arisen fully developed, in a single shot, one would indeed be entitled to invoke "irreducible complexity" explainable only by "intelligent design." But all that we know indicates that this is not what happened.

Proteins bear unmistakable evidence of *modular construction*. They consist of a number of small domains, or motifs, many of which are present in various combinations in a number of different protein molecules, indicating strongly that they have served as building blocks in some kind of combinatorial assembly process. This fact suggests that precursors of the modules at one time existed as independent peptides that carried out, in some primitive protocells, the rudimentary equivalents of the structural and catalytic functions devolved to proteins in present-day cells. This hypothesis is consistent with theoretical calculations, by the German chemist Manfred Eigen, showing that the first genes must have been very short in order for their information to survive the many errors that must have beset primitive replication systems.[8]

Granted this simple assumption, one can visualize a primitive stage in the development of life supported by short peptides and subject to a Darwinian kind of evolution in which a progressively improved set of peptides would slowly emerge by natural selection. Let one outcome of this process be enhanced fidelity of replication, so that doubling of gene size becomes possible without adverse effect. Then, random combination of existing genes could progressively generate a new set of more efficient peptides twice the length of the previous ones. Darwinian competition would once again allow progressive improvement of this set, eventually ushering in a new round of the same kind at the next level of complexity, and so on.

This stepwise model provides an answer to an objection, often associated with a call for a guiding agency and recently revived, and refined, by the American mathematician William Dembski, a leader in the modern "intelligent design" movement.[9] The objection points to the fact that life uses only an infinitesimally small part of what is known as the sequence space, that is, the number of possible sequences. Take proteins, for example. These molecules are long chains consisting of a large number of amino acids strung together. Consider such a string one hundred amino acids long, which is a small size for a protein. Since proteins are made with twenty different species of amino acids, this molecule is one among 20^{100}, or 10^{130} (one followed by 130 zeros), possible proteins of the same size. This number is so unimaginably immense that even trillions of trillions of universes could accommodate only an infinitesimally small fraction of the possible molecules. It is thus totally impossible that life could have arrived at the sequences it uses by some sort of random exploration of the sequence space, that is, by means of randomly assembled sequences subject to natural selection. Hence the claim that the choice was directed by some supernatural agency, which somehow "knew" what it was heading for. With the proposed stepwise model of protein genesis (by way of nucleic acids), which is strongly supported by what is known of protein structure, this claim becomes unnecessary. According to this model, the pathway to present-day proteins went by a succession of stages, each of which involved a sequence space that had been previously whittled down by selection to a size compatible with extensive, if not exhaustive, exploration.

Like all conjectures, the proposed model is of value only as a guide for investigations designed to test its validity. One such line of research, which I believe is being followed in some laboratories, consists in looking for enzyme-like catalytic activities in mixtures of small randomly generated peptides. Whether the model is to be rejected or whether it may continue to be entertained, possibly in amended form, will depend on the results of those and other experiments. This issue is irrelevant to the present discussion. The fact that a plausible model *can* be proposed and that its experimental testing *can* be contemplated suffices to show that we have not yet reached a stage where all attempts at explaining the genesis of complex biochemical systems in naturalistic terms have failed.

The example given illustrates the importance of biochemical knowledge for origin-of-life research. Far from discouraging such research, as Behe would have it, our newly gained understanding of the chemical complexities of life can open up valuable avenues for fruitful investigation. Today's organisms actually accomplish, by way of reactions that are now well understood and explained in naturalistic terms, what is believed to have taken place four billion years ago. They turn small organic building blocks, or even inorganic ones, into fully functional living cells. To be true, they do this within the context of existing living cells, with the help of thousands of specific enzymes that were not available on the prebiotic earth. But this fact does not necessarily impose the opinion, voiced by many origin-of-life experts, that today's metabolic pathways are totally different from those by which life first arose. On the contrary, there are strong reasons for believing that present-day metabolism arose from early prebiotic chemistry in congruent fashion, that is, by way of a direct chemical filiation. It is thus quite possible, even likely in my opinion, that today's metabolic pathways contain many recognizable traces of the reactions whereby life originated. It remains for future generations of researchers to decipher this history and conduct appropriate experiments.

Evolution

A majority of biologists subscribe in one form or another to the main tenets of the theory, first proposed by Charles Darwin, that biological evolution is the outcome of *accidentally* arising genetic variations passively screened by *natural selection* according to the ability of the variants to *survive and produce progeny under prevailing environmental conditions*. What to Darwin was largely the product of a genial intuition, backed only by observation, has received powerful support from modern biology, which has fleshed out in clear molecular terms the vague notions of hereditary continuity and variability available to Darwin. In particular, convincing proof has been obtained that naturally occurring mutations are induced by causes that are entirely unrelated, except in purely fortuitous fashion, to their evolutionary consequences. Just as Darwin postulated, blind variation comes first, with no possible eye to the future. Selection follows, enforced just as blindly by the sum total of environmental pressures, including those exerted by other living organisms.

There are a few dissenters. I don't include here the creationists and other ideologues who not only reject natural selection but deny the very occurrence of biological evolution. Neither do I have in mind the ongoing, often tedious, sometimes acrimonious debates among evolutionist coteries on such issues as gradualism, saltation, punctuated equilibrium, genetic drift, speciation, population dynamics, and other specialized aspects of the Darwinian message. The dissenters I am referring to accept evolution but reject a purely naturalistic explanation of the process.

This stand has a long and distinguished past. Especially in the French tradition, going back to the philosopher Henri Bergson, author of *L'évolution créatrice* and father of the concept of *élan vital* (vital upsurge), many biologists have embraced a teleological view of evolution, seen as directed by a special agency that somehow induced changes according to a preconceived plan. Often associated with vitalism, finalism concurrently fell into disregard with the growing successes of biochemistry and molecular biology. In recent years, the finalistic doctrine has taken advantage of the squabbles in the Darwinian camp to stage a comeback. It enjoys something of a revival in certain French circles, often with religious overtones related to the attempt by the celebrated Jesuit Pierre Teilhard de Chardin to reconcile biological science with the Catholic faith. In the Anglo-Saxon world, it lurks, in some nebulous, so-called holistic form, behind writings such as those of Lynn Margulis, world-renowned for her early—and correct—championship of the endosymbiotic origin of certain cell organelles and, more recently, for her advocacy of James Lovelock's Gaia concept and of the "autopoiesis" theory of the recently deceased Chilean biologist-cum-mystic Francisco Varela.[10]

In a different vein, the American theoretical biologist Stuart Kauffman, an expert in computer-simulated "artificial life," also expresses dissatisfaction with classical Darwinian theory. He believes that biological systems, in addition to obeying natural selection, possess a powerful intrinsic ability to "self-organize," creating "order for free."[11] The British-Australian physicist Paul Davies, best-selling author of books with titles such as *The Mind of God* and *The Fifth Miracle,* while declaring himself committed to naturalistic explanations, does not hesitate to invoke a "new type of physical law," to account for the ability of life to "circumvent what is chemically and thermodynamically 'natural.'"[12] More explicitly, finalistic views of life's origin and history are defended in two recent books, Behe's *Darwin's Black Box*, already mentioned, and *Nature's Destiny*, characteristically subtitled *How the Laws of Biology Reveal Purpose in the Universe* by the New Zealand biologist Michael Denton.[13]

The dissenters do not reject Darwinian explanations outright. They accept such explanations for many events in what is known as microevolution, or horizontal evolution in my terminology, the process whereby diversity is generated within a fundamentally unchanged body plan. They would be willing, for example, to ascribe to natural selection the diversification of the famous Galapagos finches, which Darwin found to have differently shaped beaks, adapted on each island to the kind of food available. But they deny that an unaided mechanism could possibly account for the main steps of macroevolution, or vertical evolution, which involve major changes in body plan. The transformation of a dinosaur into a bird or any other of the apparent "jumps" revealed by the fossil record belong in this category. In such cases, so the argument goes, so many important modifications had to take place simultaneously that no natural pathway involving viable intermediates is conceivable. To illustrate this point, Behe mentions the conversion of a bicycle into a motorcycle.[14] The latter can be derived from the former conceptually but not physically. However, as we have already seen, mechanical contraptions are poor models for living systems.

As with the origin of life, the issue must remain undecided, considering that the details of evolutionary pathways are unknown and unlikely to be elucidated in the near future. The problem may even be more intractable than the origin of life, since evolution is a historical process that has proceeded over several billion years, leaving very few traces. But have we reached a stage in our analysis of this process where naturalistic explanations must be abandoned and an appeal to "something else" has become mandatory? This is obviously not so. Ever since the days of Darwin, the dearth of "missing links" has been brandished as an argument against his theory. But scarcity is no proof of absence, especially in an area where the "luck of the find" plays a major role. But for *Archae-*

opteryx and the almost miraculously preserved imprint of its feathers in a Bavarian rock, discovered in 1864, no inkling as to the pathway from reptiles to birds would have been available until recently, when some findings of a similar nature were made in China. Nobody can predict what the future will yield.

Here also, modern science, far from spotlighting "irreducible complexity," may, instead, disclose unexpected ways of reducing the complexity. A dramatic breakthrough of this kind has recently occurred with the discovery of homeogenes. These are master genes that control the expression of a very large number of genes—up to 2,500, according to their discoverer, the Swiss biologist Walter Gehring, who has reviewed this fascinating field.[15] To give just one example, the so-called *eyeless* gene of the fruit fly has the power of inducing, by itself, the whole cascade of events needed for the development of a fully functional eye. Remarkably, the same gene is found in a wide variety of invertebrates and vertebrates, where it carries out the same function, even though the resulting eyes may be as different as the single-lens eyes of mammals, the similarly single-lens but differently constructed eyes of cephalopods, the multifaceted eyes of arthropods, and the primitive eyes of flatworms. Even more astonishing, the mouse gene is perfectly active when inserted into a fruit fly; but the product of this action is a typical fruit fly eye, not a mouse eye. The switch is the same; but the battery of genes that are turned on is different in the two species, resulting in the formation of entirely different eyes. Such discoveries are not only relevant to embryological development; they also illuminate evolutionary processes by showing how huge changes in phenotype can be produced by single mutations. The mysterious "jumps" may turn out to have a naturalistic explanation after all. They may even become open to experimental study.

The Elusive "Something Else"

Calling on "something else" is not only heuristically sterile, stifling research; it is conceptually awkward, at least in its modern formulation. In the days when finalism and vitalism were blended into a single, all-encompassing theory, the philosophical position had the merit of being internally consistent. A mysterious "vital force" was seen as guiding living organisms in all their manifestations. Once the stand becomes selective, accepting a naturalistic explanation for some events and denying it for others, one is faced with the almost impossible task of defining the borderline between the two. The lesson of history is that this boundary has kept on shifting, as more of the unexplained, deemed unexplainable by some, came to be explained. Many biologists are willing to extrapolate this historical course to a point in the future when all will be explained. Lest we be accused of prejudging the issue, we need not be as sanguine, simply stating that the only intellectually defensible position is to accept naturalism as a working hypothesis for the design of appropriate experiments and pointing to past successes as strongly justifying such a stand.

Behe does not accept this view. To him, the boundary is fixed, set by the limits of "irreducible complexity." On one side of the boundary, events are ruled by natural laws. On the other, there is intervention by an entity he does not hesitate to identify as God. This intervention does not stop at setting life on course and then letting it run on its own steam, so to speak. Behe's God accompanies life throughout evolution, to provide the necessary nudge whenever one is needed for some probability barrier to be overcome. Behe offers no suggestion as to the molecular nature or target of this nudge. One remains perplexed by this picture of a divine engineer creating the universe with its whole set of natural laws and then occasionally breaking the laws of his creation to achieve a special objective, which, presumably, includes the advent of humankind. Why not imagine a God who, from the start, created a world capable of giving rise to life in all its manifestations, including the human mind, by the sole exercise of natural laws?

This is what Denton postulates, but in a form perilously close to Behe's view of direct interference, in spite of its being presented as "entirely consistent with the basic naturalistic assumption of modern science."[16] Denton not only embraces the anthropic concept of a physically fine-tuned universe, uniquely fit for life, but he does so in a "hyperanthropocentric" way. He imagines the whole evolutionary script, with humankind as the crowning achievement, as written at the start into the fine print of the original DNA molecules. And he marshalls molecular arguments that purportedly support such a notion, leaving almost no room for chance events in the unfolding of evolution. He goes so far as to envisage "directed mutations," "preprogrammed genetic rearrangements," and "self-directed evolution."[17]He even admits the possibility that certain evolutionary changes may have been sustained, not by their immediate, but by their future benefits. In discussing the development of the characteristic lung of birds, he finds it "hard not to be inclined to see an element of foresight in the evolution of the avian lung, which may well have developed in primitive birds before its full utility could be exploited." Denton finds this notion "perfectly consistent with a directed model of evolution," but offers no suggestion as to how the direction could be exerted, contenting himself with the vague concept of a Creator who has "gifted organisms . . . with a limited degree of genuine creativity."[18]

A Balanced View

Naturalism has not reached the limits of its explanatory power. On the contrary, everything that has been accomplished so far encourages the belief that the origin and evolution of life are, just as are life's fundamental mechanisms, explainable in naturalistic terms. Research guided by this assumption remains a valid and promising approach to these problems. Many scientists extrapolate this scientific attitude into a philosophical worldview that denies any sort of cosmic significance to the existence of life, including its most complex manifestation to date, the human brain. Such affirmations need to be greeted with as much caution as those that claim the intervention of "something else."

Irrespective of any a priori position we may wish to entertain, one fact stands out as incontrovertible: we belong to a universe *capable* of giving rise to life and mind. Reversing a famous saying by the French biologist Jacques Monod, "The universe *was* pregnant with life; and the biosphere with man."[19] As shown by the exponents of the anthropic principle and reiterated by Denton,[20] this fact implies a considerable degree of physical "fine-tuning." Even minor changes in any of the physical constants would upset the material balance to such an extent that either there would be no universe or the existing universe would be such that, for one reason or another, life could not arise or subsist in it.

To Denton, this fact means that the universe is "designed" to harbor life. Others, however, are content with leaving it all to chance. The British chemist Peter Atkins, a militant defender of science-based atheism, calls on a "frozen fluctuation" of "extreme improbability" to account for the birth of our universe. And he concludes: "The universe can emerge out of nothing, without intervention. By chance."[21] According to the British cosmologist Martin Rees, our universe is just one—knowable because fit for life—in a huge collection of universes, a "multiverse," produced by chance.[22] The American physicist Lee Smolin has adopted the same idea, but in an evolutionary context.[23] He assumes that new universes with slightly different physical constants can arise from existing universes by way of black holes. Ability to form black holes thus appears as a measure of cosmic prolificity and serves as a selective factor in this evolutionary process. The special properties of our universe obtain because they happen to be associated with the production of a particularly high number of black holes.

Such speculations are fascinating but irrelevant to the main issue. Whether it arose by design, chance, or evolution, whether it is unique or one of a huge number, our universe *did* give rise to life and, through a long evolutionary process, to a living form endowed with the ability to apprehend, by science and also by other approaches, such as literature, art, music, philosophy, or religion, glimpses of the mysterious "ultimate reality" that hides behind the appearances accessible to our senses.[24] This fact seems to me supremely important.

In this respect, I accept the anthropic principle, but in its factual, not its teleological connotation. The universe is "such that," not necessarily "made for." I also reject the strongly anthropocentric bias given to the principle by Denton and other defenders of finalism. Humankind did not exist three million years ago, in contrast to life, which has been around more than one thousand times longer. There is no reason for believing that our species represents an evolutionary summit and will persist unchanged for the next billion years, which is the minimum time cosmologists tell us the Earth will stay fit for life. On the contrary, it seems highly probable that humankind, like all other living species, will continue to evolve, perhaps with human assistance. It is quite possible, even likely in my opinion, that forms of life endowed with mental faculties greatly superior to ours will one day arise, whether from the human twig or from some other branch of the evolutionary tree. To those beings, our rationalizations will look as rudimentary as the mental processes that guided the making of tools by the first hominids would look to us.

Another philosophical view often presented as irrefutably imposed by modern science is what I call the "gospel of contingency," the notion according to which biological evolution, including the advent of humankind, is the product of countless chance events not likely to be repeated anywhere, any time, and therefore devoid of any meaning. In the words of the American paleontologist Stephen Jay Gould, the most vocal and, because of his talent as a writer, persuasive advocate of this creed, "biology's most profound insight into human nature, status, and potential lies in the simple phrase, the embodiment of contingency."[25] I disagree with this affirmation.

What biology tells us is that humankind, like every other form of life, is the product of almost four billion years of evolutionary ramification and, put in highly schematic terms, that each new branch in this process is the consequence of an accidental genetic change that proved beneficial to the survival and proliferation of the individual concerned under the prevailing environmental conditions. The flaw in the contingency argument is to *equate fortuitousness with improbability*. Events may happen strictly by chance and still be obligatory. All that is needed is to provide them with enough opportunities to take place, relative to their probability. Toss a coin once, and the probability of its falling on a given side is 50 percent. Toss it ten times, and the odds are 99.9 percent that it will fall at least once on each side. Even a seven-digit lottery number is ensured a 99.9 percent chance of coming out if 69 million drawings are held. Lotteries don't work that way, of course; but the evolutionary game does.

The evolutionary tree has not, as is often assumed, spread out by a kind of random walk in an infinite space of variation. The space is limited by a number of inner constraints enforced by the sizes and structures of genomes and by outer constraints imposed by the environment. The numbers of participating individuals and the times available are such that widespread exploration of the mutational space has often been possible, leaving the main selective decision to environmental conditions.

In this context, contingency still plays an important role, but more so in horizontal evolution (microevolution) than in vertical evolution (macroevolution). Without leaves to provide a potential shield, no insect would have taken the weird evolutionary pathway that led it to look like a leaf—an occurrence, incidentally, that attests to the richness of the mutational field. On the other hand, if the particular historical circumstances that favored the conversion of a lungfish into a primitive amphibian had not taken place, it seems likely that vertebrates eventually would have invaded the lands on some other occasion, so rich were the selective bounties to be reaped by such a move. Nearer home, if our primate forebears had not been isolated some six million years ago—perhaps by the opening of the Great Rift Valley in East Africa— in a savannah where traits such as an erect position and manual skills became greatly advantageous, the odds are that somewhere, sooner or later, some ape group would have started on the road toward humankind. The lightning speed—in terms of evolutionary time—of hominization certainly shows that the process, once initiated, was very strongly favored by natural selection. This is not surprising, since a better brain is bound to be an asset in any environment.

Earlier in this paper, I pointed out how, by a series of successive steps of sequence lengthening and selection, emerging life could have explored substantial parts of the sequence spaces available to it at each stage. The subsequent hierarchization of the genes into an increasing number of levels—remember homeogenes—has produced something of a similar situation in later evolution, allowing widespread exploration of the relevant mutational spaces at each level of complexity. For this and other reasons, I defend the position that the vertical growth of the tree of life toward increasing complexity is strongly favored by the factors—purely naturalistic—that are believed to determine biological evolution. On the other hand, the horizontal growth of the tree toward greater diversity at each level of complexity most probably has been largely ruled by the vagaries of environmental conditions.

Conclusion

In conclusion, modern science, while increasingly doing away with the need for "something else" to explain our presence on Earth, does not by the same token enforce the view that we are no more than a wildly improbable and meaningless outcome of chance events. We are entitled to see ourselves as part of a cosmic pattern that is only beginning to reveal itself. Perhaps some day, in the distant future, better brains than ours will see the pattern more clearly. In the meantime, the stage we have reached, albeit still rudimentary, represents a true watershed in that, for the first time in the history of life on Earth, a species has arisen that is capable of understanding nature sufficiently to consciously, deliberately, and responsibly direct its future. It is to be hoped that humankind will be up to the challenge.

NOTES

1. In preparing this essay, I have benefited from the valuable advice and editorial assistance of my friend Neil Patterson.

2. See Christian de Duve, *Vital Dust: Life as a Cosmic Imperative* (New York: Basic Books, 1995; idem, "Constraints on the Origin and Evolution of Life," *Proc. Amer. Philos. Soc.* 142 (1998): 525–32; idem, *Life Evolving: Molecules, Mind, and Meaning* (New York: Oxford University Press, 2002); idem, *Singularities: Landmarks on the Pathways of Life* (Cambridge: Cambridge University Press, 2005).

3. Alexander I. Oparin, *The Origin of Life on the Earth*, 3d ed. (New York: Academic Press, 1957; first published in Russian in 1924); Stanley L. Miller, "A Production of Amino Acids under Possible Primitive Earth Conditions," *Science* (1953): 528–29.

4. Michael J. Behe, *Darwin's Black Box: The Biochemical Challenge to Evolution* (New York: The Free Press, 1996).

5. Ibid.

6. Stanley B. Prusiner, "Prions," *Les Prix Nobel 1997* (Stockholm: Norstedt Tryckeri, 1998), 268–323.

7. Behe, *Darwin's Black Box.*

8. Manfred Eigen and P. Schuster, "The Hypercycle: A Principle of Self-Organization," Part A, "Emergence of the Hypercycle," *Naturwissenschaften* 64 (1977): 541–65.

9. William A. Dembski, *The Design Inference: Eliminating Chance through Small Probabilities* (Cambridge: Cambridge University Press, 1998).

10. Lynn Margulis and D. Sagan, *What is Life?* (New York: Simon & Schuster, 1995).

11. Stuart A. Kauffman, *The Origins of Order* (Oxford: Oxford University Press, 1993); idem, *At Home in the Universe* (Oxford: Oxford University Press, 1995).

12. Paul Davies, *The Mind of God* (New York: Simon & Schuster, 1992); idem, *The Fifth Miracle: The Search for the Origin of Life* (London: Allen Lane/The Penguin Press, 1998).

13. Michael J. Denton, *Nature's Destiny: How the Laws of Biology Reveal Purpose in the Universe* (New York: The Free Press, 1998).

14. Behe, *Darwin's Black Box.*

15. Walter J. Gehring, *Master Control Genes in Development and Evolution: The Homeobox Story* (New Haven: Yale University Press, 1998).

16. Denton, *Nature's Destiny.*

17. Ibid.

18. Ibid.

19. Jacques Monod, *Chance and Necessity* (New York: Knopf, 1971).

20. Denton, *Nature's Destiny.*

21. Peter W. Atkins, *The Creation* (Oxford and New York: Freeman & Co., 1981).

22. Martin Rees, *Before the Beginning* (Reading, Mass.: Perseus Books, 1997).

23. Lee Smolin, *The Life of the Cosmos* (Oxford: Oxford University Press, 1997).

24. de Duve, *Life Evolving.*

25. Stephen Jay Gould, *Wonderful Life* (New York: Norton, 1989).

PART II

Science, Spirituality, and Society

4

Essence and Continuity of Life in the African Society

Its Evolving Nature

THOMAS ODHIAMBO

What is the purpose and meaning of human life? This, the mother of all questions, has baffled philosophers and theologians in all societies throughout the ages. Likewise, the question that naturally follows: What is the destiny of a human being within that purpose? These fundamental yet troubling questions are as relevant and perplexing today as they were during the Pharonic times of Africa, beginning more than five thousand years ago. Answers provided by sages, philosophers, theologians, and thinkers whose prime concerns are the mysteries of life, destiny, and existential continuity are still vital to society. Social institutions are built and shaped accordingly.

In contrast to the dominant contemporary worldview expressing the impersonalizing, materialistic dogma formulated by mega-giants of market societies, almost all major religions, including indigenous African religions, emphasize the evolution of human beings to a higher level of righteousness, compassion, peace, and divine insight. Righteousness and divine insight are highly personal spiritual values. Compassion and peace encompass "others"—the family, the community, the society, and the international community, the aliens, and the outcasts. The preservation of these values depends on the attention of those devoted—

through contemplation, retreat, prayer, in study and meditation—to the personal spirit within.

A tragedy in the dawn of the third millennium is that humanity, having invented unparalleled information technologies offering leisure for thoughtful discourse and unhurried contemplative study, is unable to benefit thusly. This is because human beings are frenetically occupied with political power and the elevation of "physical selfhood" to a new market utopia. All humanity is called upon to notch up the economic ratchet-wheel to the level of the modern industrialized societies of the West, regardless of their values and living conditions. There is overwhelming interest in information and knowledge, without concern for insight. Such powerful materialistic forces that consider less technically developed societies on a lower rung of cultural, scientific, and religious accomplishment are leading to the homogenization of human experience on a wholly Western neoliberal model that emphasizes the spread of consumerism and aggressive individualism. Tragically, inherent in this vision is the absence of sages, philosophers, theologians, and other intuitive thinkers whose prime vocation has been fathoming the mysteries of life, destiny, and existential continuity.

There is need now for Africans—and, for that matter, for everyone on this Earth—to give pause to these forces, to ponder again the vital questions of life's meaning. There is also need to reposition Africa on the main highway of the development of the human spirit—not as an anthropological study, nor as an apologia, but rather as a dynamically insightful plant in the grove of human experience. This chapter will consider Africa's early contribution to the evolution of human knowledge, thereafter describing contemporary African visions of the nature of God, the human spirit, and life. Questions are raised on notions of the continuity of life, nature, and human destiny, and their relevance to family and society. The traditional African worldview is highly relevant to the contemporary social and economic development process. In view of the troubling trends mentioned above, this traditional African worldview is highly relevant to the contemporary social and economic development process.

Africa's Place in the Evolution of Human Knowledge

The roots of Western religious thought may be traced to Africa, whose influence extends to Greece and underlies classic philosophy and religion. But, for the past five centuries, African history and culture have been shrouded in guilt and shame, as Maya Angelou has eloquently stated the case:

[Transported to distant lands in the Western hemisphere,] ... African slaves themselves, separated from their tribesmen and languages, forced by the lash to speak another tongue..., were unable to convey the stories of their own people, their deeds, rituals, religions, and beliefs.... Within a few generations, details of the kingdoms of Ghana and Mali and of the Songhai Empire became hazy in their minds. The Mende concepts of beauty and Ashanti idea of justice all but faded with the old family names and intricate tribal laws. The slaves too soon began to believe what their masters believed: Africa was a continent of savages.... Save for the rare scholar and the observant traveler, the Africa at home (on the continent) was seen as a caricature of nature: so it followed that Africans abroad (blacks everywhere) were better only because of their encounters with whites. Even in religious matters the African was called a mere fetishist, trusting in sticks and bones. Most failed to see the correlation between the African and his *gris-gris* (religious amulets) and the Moslem with his beads or the Catholic with his rosary.[1]

Unfortunately, the Western Enlightenment occurred at a time when mercantilist Europe became technologically capable of overseas expansionism. Without scruple, adventurers and fortune seekers began what became an unfettered conquest of Africa among other circumambient lands. Scholars provided the apologia for slavery and colonialism, sanctifying the brutalization of African lands, resources, and peoples in the drive for power, wealth, and progress. For example, David Hume wrote in his famous footnote to an essay, "On National Character": "I am apt to suspect the Negroes to be naturally inferior to the whites. There scarcely ever was a civilized nation of that complexion, nor ever any individual eminent in action or speculation, no ingenious manufacturer amongst them, no arts, no sciences." Other luminaries, including Im-

manuel Kant and G. W. F. Hegel, repeated such formulations of philosophical prejudice. Hegel positioned Africa outside history, as the absolute, nonhistorical beginning of the movement of Spirit.[2]

Maya Angelou appropriately inquires, "How, then, to explain that
these people, supposedly without a culture, could so influence the cultures of their captors and even of distant strangers with whom they have
had no contact?"[3] Myths about Africa emanating from prejudice, deliberate ignorance, and guilt's need for self-justification cannot bury the
reality of the gifts that the peoples of Africa have made to embryonic
Western civilization—gifts that were transmitted through Egypt to the
Greeks. There is growing archeological evidence and scholarly support
for the proposition that the brilliant civilization of Pharonic Egypt was
rooted in black Africa. While the intellectual contribution of Egypt to
Greek thought had never been denied, the theory that the intellectual and philosophical roots of the great Egyptian civilization may have
originated in black Africa has not been widely acknowledged, notably
in Western countries.

In 300 B.C., an Egyptian priest began to compile a history of Egypt
for the reigning monarch Ptolemy. Manetho's chronology supports the
legend that by 3100 B.C., King Menes of the Upper Nile was able to con-
quer the Delta part of the Lower Nile and establish ancient Egypt's first
dynasty. Further, according to legend, Menes ruled sixty years before being carried off by a hippopotamus. By 3000 B.C., systems of hieroglyph-
ic writings had apparently also been invented by Africans of this region.
There is considerable archeological evidence that these early Egyptians
may be descendants of an advanced black people who inhabited the wilderness of the Sahara desert for some four thousand years, during which
time, beginning in 8000 B.C., they evolved a highly organized way of
life that included the invention of the first-known calendar, presumably for predicting the rains. A shift in climate caused these highly developed people to migrate eastward into the Sudan and the Upper Nile
Valley regions, as well as toward the Horn of Africa.[4]

Along the same vein, Henry Olela argues that the modern African
worldview, as well as that of the Greeks and Romans, came from an-

cient Africans (Egyptians, Nubians). Olela, assuming Hellenic culture originated in Crete, inquired about the origins of this island's people. He found various sources suggesting that they may have been descendants of a family branch of western Ethiopians, whose origins are traced to the Sahel in 2500 B.C. He also cites other researchers who aver that the Minoan civilization had none other than an African foundation.[5] Archaeological evidence in Egypt also supports the proposition that the Egyptians were descendants of a number of African nations. So intimately intertwined are the histories of Egypt, Kush (established by the Nubians of Sudan and the upper Nile), and Axum (a civilization in the Horn of Africa) that they are inseparable in the Egyptian foundations of universal philosophical thought. Some of these peoples having tremendous impact on the Egyptians are thought to be descendants of the Gallas Somolians and the Masais of the area that is now Kenya.

In 1987, Martin Bernal's extensive study emphasizing the African roots of classic civilization was published, quoting both Herodotus and Plato as well as other sources on which he built his Black Athena theory.[6] Bernal describes the interest of the Greek historian Herodotus (450 B.C.) in how Egyptian and Phoenician colonies influenced Greek civilization—the former through the introduction of writing to the Hellenes, and the latter by their transference of the Egyptian religious mysteries, including the art of divination. The names of nearly all Greek gods were adapted from Egyptian formulations. Herodotus claimed that the names of all gods had been known in Egypt from the beginning of time. Early inhabitants of Greece offered sacrifices while praying, but their gods were neither named nor titled. The Greeks started to use the Egyptian god-names only after the Pelasgians received the approval to do so from their most ancient, and only, oracle at Dodona.

Also, according to Bernal, Plato spent time in Egypt around 390 B.C., and in his *Phaidros,* Plato has Socrates declare that the Egyptian god of wisdom, Theuth-Thoth, is the inventor of arithmetic, geometry, and writing. In his *Philebus* and *Epinomis,* Plato furnished further descriptions of Thoth as the creator of writing, language, and all the sciences. A few years later, having borrowed material from the library of

Alexandra on his sojourn in Egypt with Alexander the Great, Aristotle stated in his *Metaphysics* his belief that Egypt was the cradle of mathematics because of the toil of the Egyptian caste of priests who had the leisure to engage in such theoretical pursuits. Thus, geometry was not invented simply to measure land after the original landmarks had been washed away by the Nile's periodic floods, as Herodotus had believed.

The association between Egypt's old religion and the extraordinary rise of mathematics and science in ancient Egypt is an intriguing one. The philosophy of the old Egyptian religion was not concerned with the ephemeral and material world of "becoming"— with its cycles of growth and decay. Instead, the religion emphasized the immortal realm of being, as exemplified in numbers, geometry, and astronomy. This approach to life also influenced Greek metaphysical thought. The writings of Plato and Aristotle on these subjects are considered to have been strongly influenced by Egyptian ideas. Plato may have adopted the Egyptian view of the immortality of the soul as well as its view on creation and its doctrine of the Good. There is a strong basis for speculating that Aristotle adopted the Egyptian notion of the "unmoved mover," the creative process developed from disorder to order. This process was performed through mind and word—or pure intelligence, as Olela recounts it.[7] Aristotle may also have learned from Egypt the doctrine of the soul that he discussed in his *Book of the Dead*.

Not unlike their ancient forefathers, contemporary Africans, past and present, are extremely religious. Nearly all individual and group activities, of whatever nature—political, economic, social, or military—are heavily influenced by considerations essentially of a mystical and religious character. Their knowledge of God is expressed in proverbs, short statements, songs, prayers, myths, stories, and religious ceremonies. While there are no sacred writings in traditional societies, many Europeans traveling in Africa from the seventeenth century onward found a similar monotheistic and, in many ways, metaphysical thread of thought regarding God and the human spirit across the continent. This may be confirmed by comparing a number of reports and studies. For example, the philosophical, esoteric, and metaphysical aspects of Western African thought were reported by the French scholar Maurice Delafosse (1879–1926) in a series of monographs on indigenous religions and cultures of Africa that teased out the ritualistic and institutionalized facets from their philosophical frameworks on life and thought. Delafosse's work is part of an extensive project on African religions carried on by William Leo Hansberry.[8] Recently, John S. Mbiti has carried out a wide-ranging study of African traditional religions spanning the continent and covering some three hundred African peoples' beliefs. He has also amassed an impressive bibliography of studies on individual traditional religions.[9]

The Nature of the Supreme Being

Although concepts of God are colored by the peculiarities of the culture, geography, and environment of each nation, there are sufficiently significant elements of commonality of belief that enable one to speak of an African concept of God. It can be stated with reasonable confidence that all African cultures tend to believe that God, whatever title or name is used, transcends perception by the human faculties. While God has been associated with concrete thought forms (i.e., single objects or supernatural forces in the visible universe such as the Sun, the Moon, the high mountain, forest, or thunder and lightning), none of these figurative descriptions are incarnations of God, nor are they worshiped as gods. They are generally metaphorical representations to promote understanding of God and to enable people to draw closer to him/her. As a Dutch scholar who visited the Kingdom of Benin in West Africa wrote in an eyewitness account of 1705, "Because God is invisible, they [the citizens of Benin] say that it would be foolish to make any corporeal representations of him, for they insist that it is impossible to make an image of what one has never seen."[10]

Generally, the tangible or metaphorical descriptives of God reflect certain attributes that people associate with the invisible God and are natural touchstones to further understanding of God. For example:[11]

- The Luo of Nyanza, in Kenya, living astride the Equator on the shores of Lake Victoria, regard the Sun, or Nyasaye, as an expres-

sion of an aspect of the Supreme Being, who is the author of life
and death and the creator of all things.

- The Wachagga, who live on the slopes of Mt. Kilimanjaro, just
  south of the Equator in Tanzania, know the Supreme Being as
  Ruwa, and associate him with the Sun and the sky. They see God
  as the omnipotent creator of all things, who changes not.
- The Barotse, who live in Upper Zambezi Valley of Zambia in
  central-southern Africa, know the Supreme Being as Nambe, who
  created the universe and is the great cause of everything. He is
  omnipotent and nothing can be done against his will. The Barotse personified God through the Sun and sky, which are the more
  conspicuous abodes for their God.

From these and numerous other accounts, God possesses the eternal, intrinsic attributes of omnipotence, omniscience, and invisible omnipresence. God is the first cause and creator of the universe, that which is visible to the physical senses and that which is invisible (the visible universe
is of lesser importance than the invisible one). God is the Great Protec-
tor and Governor of All Creation. Many concrete examples throughout
Africa express these attributes:

- *Omnipotence.* There is a saying among the Yoruba of Nigeria that
  duties and challenges are as easy to do as that which God per-
  forms; as difficult to do as that which God enables not.[12]
- *Omnipresence.* The Bamum from Cameroon call God "Nnui,"
  which means "He who is everywhere." The Shilluk in the Sudan
  and the Langi of Uganda liken God to the wind and air to convey
  this attribute.
- *Omniscience.* To the Zulu of South Africa and the Banyawanda of
  Rwanda, God is known as "The Wise One," and to the Akan of
  Ghana as "He who knows and sees all."[13]

Human Nature

The relationship between God and humankind is as that of a parent and
child. God is the Father and, for some people, the Parent, or the Mother

of humanity. What then is the nature of the human being? Having a divine origin, how can human beings be considered mere flesh and bones? Generally, the human spirit and/or soul is assumed to be immaterial and divine. Each person is a complex of material and immaterial elements, the immaterial being the vital essence of the person reflected in his/her health and well-being. There are different versions of this view.[14]

According to the religious culture of the Mandingos, who live throughout the dry savannah areas of West Africa, from Niger to Senegal, the total human being is comprised of three entities: the visible, physical mortal body, which is by far the least significant entity; and two distinct, invisible, noncorporeal entities—*Dia* and *Nia*—housed in the mortal body.[15]

Mandingos consider the first noncorporeal entity, the Dia, a vital force or energy that has no will or intelligence or personality of its own. It enters the physical body at the time of conception and imparts to the body the vitality that energizes it throughout its mortal existence. The Dia is believed to be a portion of a universal and eternal form of impersonal energy that permeates the whole of nature, including inanimate as well as animate things. This universal and eternal life force springs ultimately from the Supreme Being.

The Nia, on the other hand, provides the will and the intelligence that directs the operations of the Dia as it resides in the physical body. The Nia has its own personality and, like the Dia, is believed to exist long before it enters the physical body during the uterine development of the forming infant. It is responsible for controlling and harmonizing the various materials that contribute to the development of the fetus. The Nia exists beyond the death of the physical body. During that existence, its disembodied form preserves the same moral temperament and intellectual powers that characterized it during its sojourn in the phenomenal world, as well as retaining its knowledge of and interest in the phenomenal environment and human personalities with whom it had interacted closely. It is able, for a time, to reestablish contact with them. It is clear that the striving toward spiritual perfection—the perfection of the Nia, in the case of one Mandingo, for example, is the principal purpose of human life in its totality.

Likewise, the Akan people of Ghana consider a human being to be constituted of three elements: the *okra, sunsum,* and *honam (nipadua).* The okra is the innermost self, or life force of the person expressed as a spark of the Supreme Being (*Onyame*) in the person as the child of God. It is also translated as the English equivalent of "soul." The sunsum seems to be equivalent to the spirit of humanity, and with close analysis is not essentially separate from the okra. The honam refers to the physical body. The Akan thinkers believe that there is clear interaction among these three parts. In cases of obvious bodily illness there are often times when modern Western medicine has no effect but traditional medicine is extremely effective. In such cases the bodily illness is thought to objectify a troubled spirit, so bringing it in line with God brings about healing. From the point of view of Akan metaphysics, the human and the world in general cannot be reduced to physics or an assemblage of flesh and bones. The human is a complex being that cannot be explained or satisfied in terms of a purely physical mindset concerning the universe.[16]

An essentially similar description of human nature and the purpose of life is exemplified in the religious belief system of the peoples of southern Nigeria, recorded by the English anthropologist, Arnaury Talbot (1926).[17] However, in this belief system, the human personality comprises five separate but associated entities, of which only the first is visible.[18] These are:

- The physical body—considered the least important entity of the total human being, as it is only the material, corporeal framework, through which other entities operate while resident in the earthly environment.
- The ethereal body, or the feeling self—believed to be the inner frame of the physical form. Though invisible, the feeling self consists of a material substance and, like the physical body, is subject to death on Earth, and acts as the vehicle for the vital force (or the total body's vitality) that animates the physical body during the period of earthly life.

- The mental body, or the thinking self—considered to be immortal, acts as the vehicle for human consciousness.
- The spiritual body, or the minor ego—believed to be immortal, originates from God, and is not necessarily confined to the physical body to which it is attached.
- The transcendental self, or the over-soul or the major ego—is thought to be the spark of divinity found as a kind of individualized fragment of the Supreme Being himself. It is not and never becomes an integral part of the mortal to whom it belongs but always remains with God, of whom it is a veritable and inseparable part. In its spiritually pure and perfect state, the divine spark continually transmits a stream of consciousness to the four other entities on Earth, to whom it belongs, in this way potentially enabling the latter to advance toward spiritual perfection.

This growth toward spiritual perfection is slow but gradual. The major ego keeps surveillance over the thoughts and actions of its earthly embodiment, offering to the latter rewards or punishment according to perceived good or evil. Punishment includes sickness, poverty, or other misfortunes; rewards include property, honor, and children. The major ego never prescribes a particular life style; consequently, each individual is the architect of his/her own destiny. Reincarnation is a part of the process of spiritual perfection. Following death, the soul is repeatedly sent back to Earth, to be reborn in human form to a particular station determined by its previous existence on Earth.

Continuity of Life

The concept of continuity is partially related to the notion of spirits, which, like people, are creations of God and serve different purposes. Spirits are relevant not only for immortality but also to the meaning of life on Earth and to ethics.

Among the southern Luo of Kenya, Tanzania, and Uganda, as well as among the Nilotic peoples in the Sudan and Ethiopia—the Shilluk,

Dinka, Mao, Kunama, etc.—the Supreme Being (Juok or Jok), with his supernatural power, is associated with destiny. Humans and other creatures, as well as inorganic objects, contain differing levels of supernatural power. But only humans and animals possess spirits and souls in addition to their shadows and physical bodies that decompose upon death. The spiritual attributes, the soul and spirits, continue to exist after this physical death. Humans are regarded above animals because they possess more spiritual power. Nevertheless, since animals and plants provide the resource base for human existence, they also possess some spiritual power of their own. Certain religion-ritual relationships have been established to manage this coexistence in an ethical manner.[19]

In the tradition of Luo, life is perpetual. After death, the person is said to be sent back to where he/she originated, that is, to the spiritual world, where all needs are met. In that spiritual world, some of the residents are older than others. The oldest of them all is Juok, who possesses the highest order of wisdom and is the source of supernatural power. This is the basis of the cardinal belief among the Luo peoples that age is a barometer of the accumulation of insight and wisdom.[20]

The temporal and hierarchical spiritual continuity of life has profound impact on the ethics and moral values of the Luo. Immorality jeopardizes the spiritual welfare of the entire family or village or clan, not simply that of the concerned individual (chara). Thus, souls of humans are members of a spiritual republic, dwelling for a while in the flesh. The invocation rites that an individual and the public perform, such as funeral ceremonies and sacrifices to lineage or clan, are necessary so that the community at large might be afforded insight into its condition as a spiritual entity.[21] In this context, moral strength is deeply rooted in the soul, not just in the brain.

Before concluding this discussion on African traditional beliefs, it is important to make the following observations. From the seventeenth century onward, many foreigners' observations on African religion have revealed serious misunderstandings, if not deliberate determinations to denigrate traditional thought. Notably, Charles de Brosses (1760) found fetishism to be the original foundation of all religion, in particularly in the traditions of Africa. Auguste Compte, who adopted De Brosses' theory, promulgated it with additional flourishes, notably that fetishism was the indisputable foundation of the African religious belief system and that the notion of one Supreme God in Africa was introduced by Jewish and Christian travelers.[22] These detractors and others have been part of a long history of disavowals of any central place for the African thought system that has so far not extinguished Africans' basic belief in their own religious conceptual framework.

Four of the most common misconceptions concerning ancestor worship are superstition, animism, zoolatry or paganism, and fetishism. While departed relatives are believed to live on as spirits revealing interest in their surviving families, acts of proper respect toward the departed do not mean that they are worshiped but that their wisdom and goodness should not be forgotten. Acts of remembrance and communication through prayer is only a small part of traditional religion. The vast majority of beliefs are based on deep reflection and long experience and therefore have little in common with mere superstition. Nevertheless, superstition penetrates in every religion to a degree, but this does not mean religion should be equated with superstition. While African religions acknowledge the existence of spirits, some of which are believed to inhabit physical objects, this is only a small part of religion and is to be understood in a framework of the belief that God is supreme over spirits and humankind—all are part of God's creation. Such beliefs are not the equivalency of animism or zoolatry anymore than they are of Christianity or Islam. African religion is not constructed on magic. While Africans wear religious charms, they are no more fetishists than the Christian who wears a Saint Christopher's medal or a Muslim carrying his beads.[23]

Nature and Human Destiny

In sharp contrast to the religious tradition underpinning the African worldview is the prevailing Western worldview driving the powerful forces of globalization, the modern legacy of Occidental Enlightenment thinking that enshrines aggressive individualism. While some early

Western Enlightenment thinkers denigrated what was African in their apologia for colonialism and slavery, the attention of modern thinkers is to be drawn to what, in the African tradition, is essential to the development of humane governance and the protection of the environment from the excesses of rationality, material progress, and acquisitive individualism. In other words, there is perhaps as much to be learned today from the African experience as there was in ancient times of the philosophical and metaphysical nature and meaning of life, those dimensions buried in the materialism of the spirit of the time.

It is clear that ethics in the Enlightenment tradition has focused on the ideal of the individual, while ignoring the larger human community and the wider biological and physical domains. Generally, this globally prevailing Occidental thought regards humans as distinct from the rest of nature, and not as a component of a systematic totality. This Western conception has led to environmental degradation on a broad scale. Current Occidental concerns about the environment revolve largely around two principal questions: first, whether ethics should be restricted to human beings or whether nonhuman but sentient beings should also be subjects and agents of ethics; and second, whether any other material objects in nature such as hills, rocks, and rivers possess any moral value.[24]

Indeed, in Christian doctrine, nature is usually interpreted as a bundle of resources to serve humans. This view is consistent with that articulated by Aristotle in his hierarchical justification for the ecological dominance of humans, as he asserted that plants exist for the sake of animals, while animals exist for the sake of man. Such anthropocentric, utilitarian, and hierarchical ecological insights attained the status of divine wisdom by the time of Thomas Aquinas. And, since the dawn of the modern scientific movement in the early fifteenth century, elements of nature have largely been regarded as objects for analysis and experimentation or for commercial transactions. This approach recognizes no limits to industrial exploitation, which by now has exceeded all previous limitations on the accumulation of riches and property by single individuals, corporations, or nations, since the beginning of the industrial revolution in the seventeenth century. Yet, wholesale exploitation of nature's abundance and the disruption of its intricate web of associations and relationships may well bring the entire biosphere to a sudden, grinding halt in the same manner that the intensive and unsustainable exploitation of the Fertile Crescent, the valley between the twin rivers Euphrates and Tigris, led to the sudden end of ancient Mesopotamia.

The conventional human-centered, utilitarian view of nature has not been questioned sufficiently and thus the pace of nature's destruction has not been altered. There is no doubt that a shift from rank individualism to a more compassionate universalism would provide a humane solution. Universalism approximates nature's idea of the good. It would, to a degree, oblige humanity to move away from the Occidental focus on the human individual objective as the center of ethics.

But is it feasible that such a shift could be accomplished in the lifetime of the youngest in the present spectrum of living generations, or even in the course of the next few future generations? The shift is only remotely possible, provided a monumental change can be made in several theaters of human endeavor. There is some considered guidance to be derived from African traditional religion.

Conclusion

It is clear from this brief overview of African belief systems that there is a shared cardinal belief in the Supreme Being that has been translated into a complex relational system between God and the human person. This system penetrates the whole of the education of the person with an ethic of service to the "other" and respect for nature. For example, the Luo—and this is not unique to the Luo, but shared among the indigenous peoples of the continent—have regarded the education of a person to be a lifelong experience or the sum total of experiences, which mold attitudes and govern the conduct of both the child and the adult.[25] The emphasis on the role played by the personal growth of each individual does not lead to a lessened regard for the collective. Rather, collective good hinges upon each member of the community seeing his or her actions as contributing to the well-being of the community.

Thus, for the Luo there exists a continuum between individual and collective action. This embraces education for personal growth and stresses social norms and moral ethics. Respect for age and human endeavors are guided according to values that promote unity. The elders share a common responsibility for the youth. In turn, children must respect not only their parents but all who undertake responsibility for them, including members of the lineage, clan, and tribe. With this method of education, young people accomplish many things. This method strengthens kinship ties throughout the particular ethnic group; provides an education for family life; instructs children about the clan's religion; and teaches the normative rules of social conduct. It is interesting to note that a prison system was unheard of among the Luo peoples. Professional education operates on a different level, being confined to selected people, sometimes within a single clan, whose skills and knowledge are transmitted by a strict code of apprenticeship from one generation to the next. These professions, including blacksmithing, medical practice, or theology, again have the well-being of the other in focus. Possessive individualism is a behavior that was—and still is—abhorred in the traditional African society.

From the perspective of this African worldview, global measures of human progress should be more inclusive. It must extend beyond gross national product, expenditure on the armed forces, and other similar comparative national statistics. And a modified ranking should take into consideration other perceptions of progress—not necessarily quantifiable—in the field of social and community relationships. These include the practice of personal moral values; love for one's family, neighbors, and strangers; attention to community goals; compassion for the economically poor, the orphans, and the widows; and the use of incentives and social sanctions against offenders of social and community norms rather than imprisonment and execution. These factors are also essential signs of human progress.

The foundation for this wider and more humane vision of progress would be firmly built if based on a continuous search for divine insight, central belief in a supernatural order beyond the physical world, and a deep sense of humility and tolerance for another's views. In international relations, this would include an integral belief in peace, evolution beyond the need for the political and military domination of other nations, and the progressive adoption of a thought system that recognizes no chosen people or superior race. Such a repositioning of Africa in the grove of human experience and the role it can play in the project on the human spirit are matters that require deep contemplation, prayer, and study in retreat from the clamor for wealth, power, and prestige that crush the modern human spirit.

NOTES

1. Maya Angelou, *Even the Stars Look Lonesome* (New York: Bantam Books, 1997), 15–16.

2. E. Chukwudi Eze, "Modern Western Philosophy and African Colonialism," in *African Philosophy*, ed. E. Chukwudi Eze (Oxford: Blackwell Publishers, 1998), 214.

3. Angelou, *Even the Stars*, 16.

4. Similar accounts are offered in the following sources: John Lanphear and Toyin Falola, "Aspects of Early African History," in *Africa 3*, ed. Phyllis M. Martin and Patrick O'Meara (Indianapolis: Indiana University Press, 1995), 73–75; Vivian Davies and Renee Friedman, *Egypt* (London: British Museum Press, 1998), 16–28.

5. Henry Olela, "The African Foundations of Greek Philosophy," in *African Philosophy*, ed. E. Chukwudi Eze (Oxford: Blackwell Publishers, 1998), 43–49. Olela cites William N. Huggins and John G. Jackson, *Introduction to African Civilization* (New York: Negro University Press, 1969), 77; Leonard Cottrel, *The Penguin Book of Lost Worlds* (London: Penguin, 1966), 24.

6. Martin Bernal, "Black Athena: The Afroasiatic Root of Classical Civilization," in *The Fabrication of Ancient Greece 1785–1985* (New Brunswick: Rutgers University Press, 1987), 98–99, 106–9, 49–52.

7. Olela, "The African Foundations of Greek Philosophy," 48–49.

8. William Hansberry, "Indigenous African Religions," in *Africa from the Point of View of American Negro Scholars*, ed. J. A. Davis (Paris: Presence Africaine, 1958), 89.

9. John S. Mbiti, *African Religions and Philosophy* (Oxford: Heinemann Press, 1969), 81.

10. William Hansberry quotes David Von Nyendaci in "Indigenous African Religions," 85.

11. Ibid., 89–90.

12. E. B. Idowu, *Olodumare: God in Yoruba Beliefs* (London: Longman Press, 1962), 40–41.

13. Mbiti, *African Religions and Philosophy*, 31.

14. E. W. Smith, *Doctrine of God* (London: Lutterworth Press, 1944), 55.

15. Maurice Delafosse, "Souffle vital et esprit dynamique chez les populations indigenes du Soudan occidental," *Comptes-rendues des seances de l'Institute Francais d'Anthro-*

pologie, Supplement No. 5 (1912); and idem, *Croyances et pratique religieuses, Les Noirs de l'Afrique,* chap. 9 (1922). Cited in Hansberry, "Indigenous African Religions," 90–96.

16. Kwame Gyeke, "The Relation of Okra (Soul) and Honam (Body): An Akan Conception," in *African Philosophy,* ed. Eze, 59–65.

17. P. A. Talbot, *The Peoples of Southern Nigeria* (London: Oxford University, 1926), 93–96, as cited in Hansberry, "Indigenous African Religions."

18. See, for example, John S. Mbiti, *Introduction to African Religions,* 2d ed. (London: Clay Ltd., St Ives, 1975), 70–82.

19. A. B. C. Ocholla-Ayayo, *Traditional Ideology and Ethics among the Southern Luo* (Uppsala: Scandinavian Institute of African Studies, 1976), 191–200.

20. Ibid., 172–74.

21. Ibid., 55–56.

22. See, for example, Charles de Brosses, *Du Culte des dieux Fiches,* 1760, quoted in Hansberry, "Indigenous African Religions," 97–98. See also L. Wilson, *Western Africa: Its History, Condition, and Prospects* (New York: Harper and Bros., 1858), 220.

23. Mbiti, *Introduction to African Religions,* 17–20.

24. O. Oruka and C. Juma, "Ecophilosophical and Parental Earth Ethics (On the complex Web of Beings)," in *Philosophy, Humanity, and Ecology,* vol. 1: *Philosophy of Nature and Environmental Ethics,* ed. H. O. Oruka (Nairobi: Acts Press and The African Academy of Sciences, 1994), 1:15.

25. Ocholla-Ayayo, *Traditional Ideology and Ethics,* 58–60.

5

Einstein and Gandhi
The Meaning of Life

RAMANATH COWSIK

Motivated by the extraordinary lives and thoughts of Einstein and Gandhi, the aim of this chapter is to show that science and spirituality provide us with complementary perspectives on truth, both unbiased and universal. Such a perspective motivates us to realize the futility of human desires and mundane passions and to develop a feeling of universal empathy, thus inducing us to work for the betterment of the world. It is this selfless toil that imbues life with meaning.

Einstein's scientific contributions revolutionized almost every aspect of modern physics: quantum theory, theory of space-time, gravitational physics, and statistical physics. He redefined the very concept of space-time, in which physical events take place, and the objective reality in quantum systems. Whereas the Copernican revolution that started nearly five hundred years ago moved us away from a geocentric point of view, Einstein's theories of relativity connected space and time into a single manifold and made the very question as to where the center of the universe lies meaningless; there is the absolute freedom of choice. Moreover, the equations of Einstein's theory of gravitation revolutionized cosmology in the following way.

The Earth on which we live is about 150 million kilometers away from the Sun, which is a star. Of the stars that fill the firmament, about

100 billion of them are conglomerated as the Milky Way Galaxy. Again, there are scores of billions of galaxies distributed through space in a quasi-random way. Thus, the cosmological principle states that the universe is homogeneous and isotropic on large scales. When Einstein's equations were used to investigate the consequences of this aspect of the universe, the solutions indicated that the universe was expanding in a very special way: the galaxies were moving apart from one another, similar to dots on an expanding balloon. It was as if the fabric of space was being created and causing the distance between the galaxies to increase. In 1924, Edwin Hubble established that the galaxies were indeed moving away as predicted by Einstein's equations.

Astronomical research during these intervening years has shown that the universe has indeed expanded from an extremely hot condensed state called the big bang. As the universe expanded and cooled, the primordial exotic particles and fields gave birth to the quark-gluon plasma, familiar in the context of the Standard Model of particle physics today. By the time the universe was just one second old, the quarks had combined and we had the particles of nuclear physics: neutrons, protons, electrons, positrons, neutrinos and neutrino-like particles, and, of course, radiation. But these particles were still too hot. At five minutes, the universe had cooled enough to synthesize helium. Since the nucleus with mass 8 is unstable, helium nuclei could not sequentially fuse to provide heavier nuclei, and so the universe consisted of neutrinos and neutrino-like particles, electrons, protons, α-particles, or nuclei of helium and radiation; there was no carbon, nitrogen, oxygen, iron, or heavier elements. The building blocks of life and our familiar world were yet to be made.

For a million years, the universe went through an uneventful expansion, just cooling down continuously. Eventually, the temperatures were cool enough for the electrons and protons to combine to form atoms of hydrogen, which suddenly released the close coupling between radiation and matter, with dramatic effects. During the cooling-down process, the neutrino-like particles had also cooled and their random motions slowed down, allowing their self-gravitation to clump them together into clouds. Since these neutrino-like particles neither emit nor scatter light, they are called dark matter particles. The clouds of dark matter gravitationally attracted the atoms, which radiated and slowly settled into the central regions of the clouds. Such clouds with atomic gas merged to form galaxies. Our Milky Way is one such system.

The gas in the central regions in such systems condenses into stars. The central core of a star has a temperature of about 10 million degrees Kelvin, and here nuclei of hydrogen and helium fuse to form the heavier elements, which are then dispersed back into the interstellar space by stellar winds. Occasionally, when the mass of the stellar core exceeds the Chandrasekhar mass, it undergoes a collapse under self-gravity and the outer regions are expelled in an explosion; this debris contains the heaviest elements, even up to uranium. In the 8 to 10 billion years since the birth of the universe, such processes seeded most of the galaxies with heavy elements. Our solar system formed in one such galaxy about 4.6 billion years ago. Thus, everything that we see about us has an intimate connection with the birth of the universe and with the subsequent stages of its evolution. We are all made of stardust.

It is only during the last few billion years that life appeared on this earth in the form of unicellular organisms, and the slow evolution of the species led, finally, within the last hundred thousand years, to humankind as we know it. The history of civilized man with agricultural capabilities is even shorter—a mere ten thousand years or so.

Two points may be noted here: first, a systematic and progressive sequence of evolution has brought the world to its present state. Man himself, with his intelligence and capacity for articulation and organization, is shaped by the progressive evolution of the exotic particles and fields of the early universe—the formation of galaxies, nucleosynthesis in the stars, the origins of life on this planet, and its subsequent evolution into modern man. The second point that should be underscored is that the span of man's existence is but a miniscule speck in this vast universe, which is about 14 billion years old and has an extent of 10^{23} km. Yet, man's indomitable spirit has striven to comprehend this cosmos. I shall return to the discussion of these two points shortly.

Continuing with our description of Einsteinian cosmology, we note that normal matter such as hydrogen and helium contribute only about 2 percent to the average mass density of the universe. In contrast, the neutrino-like particles of dark matter contribute about one-third of the mass density on the average, and these dominate the formation and dynamics of the galaxies. What is the remaining 65 percent made of?

Einstein, at the time of inventing the relativistic cosmologies, had also discovered a way of causing the expansion to be either halted or accelerated. In a manner that was in perfect consonance with the mathematical aesthetics of physics, he had introduced the Λ-term into his field equations. Such a term finds support in the concept of the "quantum-vacuum," according to which even perfectly empty space has a dynamic of its own—with particles, antiparticles, and radiation continuously being created and annihilated, all in a manner perfectly in agreement with the conservation of energy and quantum mechanics. It is remarkable that during the last ten years astronomical evidence is mounting that such a vacuum or dark energy indeed is present, and accounts for the 65 percent missing energy density. It shows its unique vacuum character, i.e., of a gravity that repels, by making the universe accelerate in its expansion! Thus we see that normal matter, which we are all made of, is only a tiny fraction of dark matter. Furthermore, the dynamics of the universe now is controlled by vacuum energy, which is not matter at all. All this reinforces our connectivity with the universe and at the same time leads us away from a simple anthropocentric view.

Let us now briefly turn to Einstein's spirituality. His god-concept was more sophisticated than the common view of a personalized God who is the lawmaker, punishing man for his sins and rewarding him for his virtues. He said, "My comprehension of God comes from the deeply felt conviction of a superior intelligence that reveals itself in the knowable world."[1] His religion was an attitude of cosmic awe and a devout humility before the harmony in nature. Einstein considered himself an agnostic and his spirituality was closely similar to that taught by Buddha and much later by Spinoza, not unlike the *paramarthika* or the transcendental interpretation of the Vedanta delineated by Shankara in contrast to the *vyavaharika* view held by the common man. In close parallel with the Hindu saints, especially Gautama Buddha and Shankara, he felt the futility of human desires. Individual existence in the pursuit of mundane materialistic goals impressed Einstein as a sort of prison, and he felt a deep inner urge to experience the universe as a significant whole.

Thus, Einstein's spirituality is close to the philosophy of *advaita* of Shankara. Just as Einstein opened up science, which had reached a watershed in the beginning of the twentieth century, so did Shankara revitalize the religions of India with spirituality in the sixth century. Einstein felt that whatever there is of God and goodness, it must work itself out and express itself through us; we cannot stand aside and "let God do it." He was truly a *karmayogi* and followed the dictum of Gita *"Maté sangostvakarmani"* (Do not detach yourself from your duty), as he strove incessantly to prevent war and bring peace among the nations.

It should be emphasized that there is universality to Einstein's cosmic experience that is closely akin to that of the monks and nuns in deep and fervent prayer or of the mystics of the East during meditation. A common characteristic is that these experiences are so intense that they transform the individual in a fundamental way. The neuroscientist Andrew Newberg has pointed out that these "religious" experiences are common to all faiths; they induce a sense of oneness with the universe and a feeling of awe that infuse such experiences with great importance.[2] Devotees feel their sense of self dissolve, they feel a loss of boundary, and their sensory inputs weaken and even turn off completely. The attendant psychosomatic reactions imbue such experiences with deep significance characterized by great joy and harmony—similar to the feeling of parents when they see their newborn offspring—a feeling described as *bhakti* by the spiritual leaders of India. A part of the nervous system of creatures, including humans, has been perhaps hardwired this way to ensure the survival of the species and sustain evolution.

Jean Staune, Philip Clayton, and other authors for this volume have noted that a shroud of disenchantment progressively covers all of us as science describes both humankind and nature in purely reductionist terms, somehow depriving life of meaning and values. Ever since Descartes and Locke made their powerful and important contributions, the theory of knowledge has progressively banished considerations of values from their central place in human thought. It is fair to say that over the recent decades, the discussion of values has again been taken up so as to provide the foundations for ethical and moral systems. One of the distinguishing features of Indian philosophy is the continual unwavering importance attached to the discussion of values, a characteristic preserved over time, perhaps because the barriers of distance and language from Europe prevented an overemphasis of the reductionist paradigm.

Postponing a detailed discussion of the Indian values to a later occasion, let us focus attention on the implications of Einsteinian cosmology to the question at hand. The two points that were underscored during the discussion of cosmology are: (1) our connectivity with the grandest events in the universe and even to the big bang, through a sequence of evolution; and (2) the extremely miniscule span of humans in the vastness and enormity of cosmic space and time. Even this earth upon which we live is more than four billion years old—an enormous expanse of time compared with man's sojourn on it. The subtlest conditions of light, heat, water, and a proper mix of the elements led to the birth of life on this planet about three billion years ago. During most of the epochs of evolution, nature was all powerful. Nature nurtured life and made life forms that became progressively stronger, and then man appeared on the scene. He, too, was nurtured by nature, and even though he is but nature's creature now, for the first time, he has become so powerful that he can control Mother Nature. He can choose to destroy her or he can protect her and make her even more beautiful. Science alone cannot and will not tell us what we should do. Spirituality has a prescription but cannot adequately defend it. But a complete perspective jointly provided by science and spirituality can point to a set of values that will guide us to make the right choices.

Let us for a moment take inspiration from our connectivity with the rest of the universe and sensitize ourselves to the character of progressive evolution to higher levels that is innate in us. To assume that those values that support such an evolution are the right ones is both natural and consistent with the teachings of the great leaders of mankind such as Buddha, Jesus, and Shankara. When we recognize our connectivity with the rest of the world—with the inanimate mountains, deserts, rivers, and the oceans and the living things upon this earth, the trees, grass, and flowers of every hue and birds and animals including man—and we sensitize ourselves to our common origins, we will be endowed with an empathy that will give us strength to follow the precept of universal love, including "love thy enemy," as taught by Jesus among others. Is this really possible or is it just an ideal that we cannot reach?

Mohandas Karamchand Gandhi, with his deep commitment to *ahimsa* (nonviolence) and *satyagraha* (pursuit of truth), showed that one could live by the precept of the Christ. Speaking of Mahatma Gandhi and the peaceful movement he launched in South Africa and India to gain freedom from prejudice and oppression, Einstein said,

A leader of his people, unsupported by any outward authority; a politician whose success rests not upon craft nor on mastery of technical devices; but simply on the convincing power of his personality; a victorious fighter who has always scorned the use of force; a man of wisdom and humility; armed with resolve and inflexible consistency, who has devoted all his strength to the uplifting of his people and the betterment of their lot; a man who has confronted brutality with the dignity of a simple human being, and thus at all times risen superior. Generations to come, it may be, will scarce believe that such a one as this ever in flesh and blood walked upon this earth.[3]

Volumes have been written about Gandhi. The quotation from Einstein touches upon some of the salient aspects of his personality. Let me merely add that Gandhi was born in India, about ten years before Einstein, and discovered the method of peaceful non-cooperation in South Africa. This method of bringing about sociopolitical change peacefully through moral persuasion rather than through the use of force is called *satyagraha*. Even General Jan Christian Smuts, who always exerted iron-handed control, is said to have remarked,

I do not like your people and I do not care to assist them at all. But what am I to do? You help us in our days of need. How can we lay hands upon you? I often wish you took to violence like the English strikers and then we would know at once how to dispose of you. But you will not injure even the enemy. You desire victory by self-suffering alone and never transgress your self-imposed limits of courtesy and chivalry. And that is what reduces us to sheer helplessness.[4]

As the quintessence of Gandhi's virtues, I may perhaps state universal love, *ahimsa* (nonviolence), and *satya* (truth). These three qualities blend in him, supporting and adding glory to one another. These qualities became luminously clear during the long struggle for freedom in India. The unflinching and unwavering adherence to truth, not unlike that of an exemplary scientist, is at the heart of his personality, a quality from which emerge his Christ-like love and his nonviolence even in thought. In support of this idea, we may quote Gandhi himself:

To see the universal truth face to face one must be able to love the meanest creation as oneself. . . . For me the road to salvation lies through incessant toil in the service of my country and humanity. In the language of the Gita, I want to live in peace with both friend and foe.[5]

Thus, not surprisingly, he called his freedom struggle *satyagraha*, or pursuit of truth. This method proved remarkably successful, time and again, in bringing freedom from discriminatory control of one people by another—a freedom that was permanent and that left both the people not in antagonism but in friendship. Thus we see the two facets of Gandhi's personality: the spiritual inner self forever devoted to the pursuit of truth and the outer self that found expression in this world in his deep love of humanity and in his untiring efforts toward its betterment.

Apart from these personal qualities that helped Gandhi face fearlessly any onslaught, including incarceration, during his *satyagraha* movement, he had another deep idea that has relevance even today. He felt that no individual, no group or nation, whether poor or rich, should be without gainful employment. Just as the poorest eking out a living can be redeemed when provided with an opportunity to work and earn a living, even the rich, either through inheritance or in a nation with easily accessible mineral deposits, benefit greatly if they work hard regularly in their chosen fields of interest. The *charka* or *khadi* program of Gandhi was a tremendous help to the poor in India in the 1930s. Even today no one can remain merely a consumer. All of us should be engrossed in creative effort—this will give "meaning" to our lives.

Impressed by Gandhi's Christian love and indefatigable energy, Romain Rolland describes Mahatma Gandhi as the "St. Paul of our days."[6] Equally impressed by his frugality and asceticism and his self-identification with the poorest of the poor, C. F. Andrews aptly likens him to Saint Francis of Asissi.[7] To use the words of Martin Luther King:

There is another reason why you should love your enemies, and that is because hate distorts the personality of the hater. There is a power in love that our world has not discovered yet. Jesus discovered it centuries ago. Mahatma Gandhi discovered it a few years ago, but most men and women never discover it. They believe in an eye for an eye and a tooth for a tooth; but Jesus comes to us and says "this isn't the way."[8]

Nelson Mandela was also inspired by Gandhi and, in a remarkable achievement, he brought apartheid government in South Africa to an end and established universal democracy. Both King and Mandela were individually awarded Nobel Prizes in profound recognition of the idea that the Gandhian method of nonviolence is the answer to the crucial political and moral questions of our times and the need of the hour is to overcome our fears and move courageously toward peace along the Gandhian path.

Thus we see that science and spirituality both tell us that we should work to sustain the positive universal evolution or, in other words, follow *dharma* according to the Hindu scriptures. And in our incessant effort toward peace—which is essential for this positive evolution— we should follow the path shown by Buddha, Jesus, and Gandhi. This method is not restricted to the oppressed and the poor but is open to the rich and powerful as well, as indeed Asoka the Great showed more than two thousand years ago.

To summarize, we see that the reductionist approach of science has clearly pointed out our connectivity with the rest of this vast universe and events that occurred in the depths of time. Science has also shown that a positive vector of evolution has transformed the exotic fields and particles of the big bang into the world in which we live. But the reductionist approach, as it stands today, cannot tell us how to attach value to things or actions. We can resolve this impasse by augmenting the reductionist approach with an additional axiom. Let us say that all actions and attributes that support the positive evolution we referred to have a positive value. For example, love of humanity, nonviolence, and efforts toward betterment of the world will now be endowed with positive value, just as the great spiritual leaders have been telling us all along. But their message could not find support in minds rigorously trained in the reductionist approach, which tended to ignore the subtle urgings of the inner self. This extra axiom allows us to bridge the gap between science and spirituality and gives meaning to lives dedicated to bringing about peace and tranquility in this world and to lives engaged in the creation of beautiful art, sensitive poetry, and, yes, to lives engrossed in science bringing us ever closer to truth. I cannot do better to end this chapter than by quoting Rabindranath Tagore.[9]

> Where the mind is without fear and the head is held high;
> Where knowledge is free;
> Where the world has not been broken up into fragments by narrow
>   domestic walls;
> Where the words come out from the depths of truth;
> Where tireless striving stretches its arms towards perfection;
> Where the clear stream of reason has not lost its way
>   in the dreary desert sand of dead habit;
> Where the mind is led by thee into ever widening thought and action—
>   into that heaven of freedom, my Father, let my country awake.

NOTES

1. Alice Calaprice, ed., *The Expanded Quotable Einstein* (Princeton: Princeton University Press, 2000), 223.

2. For example, Andrew B. Newberg and Eugene G. d'Aquili, *The Mystical Mind* (Minneapolis: Fortress Press, 1999).

3. Written for the occasion of Gandhi's seventieth birthday in 1939, "Mahatma Gandhi" published in Albert Einstein, *Out of My Later Years* (New York: Philosophical Library, 1950).

4. C. F. Andrews, ed., *Mahatma Gandhi: His Own Story* (New York: The Macmillan Company, 1931), 247.

5. Ibid., 353–54, 357.

6. Romain Rolland quoted by C. F. Andrews in "The Tribute of a Friend" in S. Radhakrishan, *Mahatma Gandhi: Essays and Reflections of His Life and Work: Presented to Him on His Seventieth Birthday*, October 2, 1939 (London: George Allen & Unwin, 1939), 49.

7. Ibid., 50.

8. Martin Luther King Jr., "Loving Your Enemies" (sermon, Dexter Avenue Baptist Church, Montgomery, AL, November, 17 1957) from MLK Papers Project Sermons in "A Knock At Midnight," http://www.stanford.edu/group/King/publications/Sermons/571117.00.

9. Rabindranath Tagore, "Song 35" in *Gitanjali* (London: Chiswick Press, 1912), 18.

[6]

Dialogue of Civilizations

Making History through
a New World Vision

AHMED ZEWAIL

The 2002 UNESCO conference, "Science et la quête du sens" in Paris, was devoted to science and the quest for meaning; the English title, "Science and the Spiritual Quest," emphasizes the spiritual dimension, a realm beyond science. Similarly, this chapter,[1] which is based on my lecture given at the conference, is concerned with dimensions beyond science—our human existence in civilizations and cultures that may or may not be in a state of clash. As a scientist, I find these issues complex, but it is precisely this complexity that necessitates a new nondogmatic and rational approach in our quest for human understanding, our search for the truth and new knowledge through science, and our comprehension of the meaning and value of life through faith. My thoughts and reflections are guided by my experience so far in at least three civilizations—the Egyptian, the Muslim-Arab, and the American.

In thinking about the new century and the emerging world, some intellectuals have introduced concepts such as the "clash of civilizations," as termed by Samuel Huntington, and the "end of history," as expressed by Francis Fukuyama.[2] Both authors argue their cases with conviction; nonetheless, these ideas are controversial and debatable. As a scientist, I find no "fundamental physics" to these concepts. In other words, it is not a fundamental principle of civilizations that they be in a state of clash with each other. Neither is it a fundamental principle to end history with one system over all other ideologies.

Here, I argue that the current world disorder results in part from ignorance about civilizations—lack of awareness or selective memory of the past and lack of perspective for the future—and in part from the economic misery and political injustices experienced by the have-nots, which represent some 80 percent of the world's population all across the globe and in *different* civilizations. These are the barriers for achieving the advanced state of world order and, if we can overcome them, we will reach the optimum—a dialogue of civilizations.

Dialogue or Clash?

According to the dictionary, civilization means an *advanced state* of human society in which a high level of culture, science, industry, and government has been reached. Individually, we are civilized when we reach the advanced state of being able to communicate with and respect others of different customs, cultures, and religions. Collectively, we speak of globalization as a means for bringing about prosperity in the world, yet globalization cannot be a practical concept if there are clashes of civilizations. Historically, there are many examples of civilizations that have coexisted without significant clashes.

The central argument in Huntington's thesis is that in this post–Cold War era, the most important distinctions among peoples are not ideological, political, or economic but cultural. He emphasizes the point that people define themselves in terms of ancestry, religion, language, history, values, customs, and institutions; he divides the world into eight major civilizations: Western, Orthodox, Chinese, Japanese, Muslim, Hindu, Latin American, and African.

I have several difficulties with this analysis, and perhaps the following questions and commentary may clarify my position. First, *What is the basis for these divisions of civilizations?* People belong to different cultures, nations have experienced (and continue to experience) different cultures, and nations on the same continent may be influenced by different civilizations. In my case, from birth to the present time, I can identify myself as Egyptian, Arab, Muslim, African, Asian, Middle Eastern, Mediterranean, and American. Looking closely at just one of these civilizations, I note that the Egyptian people belong to a dynamic civilization with a multicultural heritage: Pharaonic, Coptic, Arabic, Islamic, not to mention the Persian, Hellenistic, Roman, and Ottoman influences. The same can be said of the European and American civilizations and others on different continents. The Western cultures of Europe, the United States, and Australia are far from uniform and homogeneous. Given the number of cultures within Europe and the United States, we should then expect a clash of civilizations within a single civilization, without having to look to the other seven. The forces uniting cultures and civilizations are not the result of simple divisions.

A second question is, *Is it fundamental that differences in cultures necessarily produce clashes?* Huntington contends that if the United States loses its European heritage (English language, Christian religion, and Protestant ethics) and its political creed (e.g., liberty, equality), its future will be endangered. From a personal point of view, I did not speak English when I came to the United States; I am not a Christian; and I was not taught Protestant ethics. Yet I integrated myself into my new, American culture while preserving my native culture(s), and I believe that both my "Eastern" and "Western" cultures have benefited from the marriage, without a clash. From a broader perspective, America's strength has traditionally been in its "melting pot"; the country has been enriched—and continues to be enriched—by multi-ethnicity and the different cultures of its inhabitants. As a result, tolerance for different religions and cultures has become part of the American civilization. Provided the people can live in a constitutionally sound system of liberty and equality, intranational clashes are not fundamental; other problems are.

Turning to international relations, it is not obvious to me why civilizations have to acquire their power through imperialism at the expense of the others. Cultures and civilizations can be at their peak of achievement and yet coexist in harmony and even complement each other. The United States, Japan, and European nations are examples of this beneficial coexistence, created by building economic and cultural bridges. The key for achieving this state is to be part of a cooperative world system that represents and observes human liberty and fairness, and whose resolutions are enforced and implemented in a timely manner. This is difficult to achieve, granted, but I believe that visionary leadership can bring it within reach.

A final question is, *What about the dynamics of cultures?* Cultures are not static; they all change with time, and the degree of change is governed largely by forces of politics and economics. Let us consider my home country. Egypt's civilization was developed very early in human history and dominated the world for millennia, but lately the nation has become a developing one. This does not mean that Egypt has lost its civilization, but it does mean that, like others, it has changed with time due to many internal and external forces. In other words, the current state is not due to some intrinsic human or genetic flaw, but rather to the changing fortunes of time.

Other examples of cultural change in Europe and other parts of the world are well known, but the dynamics may be different—different in their time scales and the forces that provoke change. In all cases, however, the dynamics of change cannot be attributed solely to the intrinsic values of an isolated culture. We must take into account political and economic interactions within a culture and between the various cultures of the world. For example, the people of North and South Korea are of similar culture, but the notable disparity in progress between the two countries is due to economic and political factors; the same can be said of East and West Germany before reunification.

The above commentary does not address a problem that is fundamental and common to all cultures and civilizations—the population of have-nots, who have a dynamic of their own. During the European Middle Ages, the peak of Islamic civilization, the majority of Europeans were have-nots, but now most nations of the Muslim world are developing or are underdeveloped, with large populations of have-nots. Some may believe that this is due to a flaw in the intrinsic values of the religion of Islam. It may be useful for me, as an educated Muslim (although I am not a scholar of Islam), to highlight some of the misunderstood principles of Islam and its dynamic civilization. This is also timely given the tragic events of September 11 (2001) in New York and Washington, D.C., their aftermath, and the association in many people's minds of these events with Islam.

Islam and Its Foundations

What is Islam? Islam is the religion and the way of life of about one-fifth of the world's population. There are 1.3 billion Muslims in the world today, 20 percent of whom are Arabs; 5 percent of Arabs are not Muslims. In 1970, there were 500,000 Muslims in the United States; now there are 6 to 7 million, 23 percent of whom are U.S. born. *Islam* is an Arabic word with a double connotation: "peace" and "submission to the will of God." Islam considers itself to be the continuation and the culmination of the earlier "God-sent" religions, Judaism and Christianity; the three are commonly called the monotheistic Abrahamic religions. God commands Muslims to respect all humanity, and Jews and Christians are referred to with distinction as the People of the Book, since they are fellow worshipers of the one God and the recipients of his scriptures (the Torah through Moses and the Gospel through Jesus). The prophet of Islam is Muhammad, who also is the descendant of Abraham through his first son, Ishmael.

Two concepts are basic in Islam[3]: the concept of the unity of God, and the concept of Islam as a way of life, including the civil and legal system. These two concepts are the core of the creed. The Islamic codes of morality are similar to those found in Christianity and Judaism. Muslims accept five primary obligations, commonly called the "five pillars" (*arkan*) of Islam. In practice, of course, Muslims can be seen observing them to varying degrees, for the responsibility of fulfilling the obligations lies on the shoulders of each individual. The pillars are the profession of faith (*shahadah*), prayer (*salah*), almsgiving (*zakah*), fast-ing (*sawm*) during the holy month of Ramadan, and performance of the pilgrimage (*hajj*), the journey to Mecca, for those who can physically and materially afford it, at least once in one's lifetime. Muslims also accept *shariah*, the body of Islamic sacred laws derived from the *sunnah* (custom and religious practice of the Prophet), the *hadith* (documented sayings and teachings of the Prophet), and the Qur'an.

The Qur'an is the scripture of Islam, and Muslims believe it to be authored by God himself and revealed to Muhammad by the Angel Gabriel. The word for God in Arabic is "Allah" and it is used by all Arabs, even Arab Christians and Jews. The Qur'an was revealed in segments of varying length, addressing various issues and circumstances, over the span of twenty-three years, the period of Muhammad's prophethood. Because it is God's direct words, the Qur'an remains in its original language, word for word and letter for letter. Once rendered into any other form or language (even Arabic), it is no longer called the Qur'an, because the direct, divine words are replaced by human words, called interpretations or translations of the meaning. The literary style of the Qur'an is so powerful that to the early Arabs it was an inimitable miracle. The style appears to share features with poetry—again, the Qur'an defies description, being considered neither poetry nor prose but a class unto itself—and this poses difficulty for some non-Muslim readers who like Bible stories told in chronological order. There is one story in the Qur'an (of Joseph) that unfolds chronologically, and to these readers it may still seem poetic.

The Qur'an makes explicit statements about human existence, integrity, and on everything from science and knowledge to birth and death. "Read!" is the first word in the first verse of the direct revelation to the Prophet (Surat al-'Alaq 96:1), and there are numerous verses regarding the importance of knowledge, science, and learning; Muslims position scientists along with the prophets in the respect they are due. The Qur'an provides a general call to humanity: "Cooperate with one another in righteousness and piety, and do not cooperate in sin and transgression" (Surat al-Ma'ida 5:2).

Tragically, some fanatics and some in the media abuse Islam and distort the meaning of its principles through terms such as *jihad* and *terrorism*. The word jihad, for example, is now routinely translated as "holy war," specifically the kind of holy war practiced by Muslims against unbelievers or infidels. This phraseology is far removed from the true concept of jihad in Islam. According to *Lisan al-'Arab*, the most authoritative Arabic dictionary, the word jihad, which derives from the root verb *jahada*, means simply to exert *maximum* effort or striving. The theological connotation of this maximum effort is that it is exerted in striving for betterment—in the struggle within oneself for self-improvement, elevation, purification, and enlightenment. For example, in Egypt, the word *mujtahid* as applied to students means a high achiever. There are other forms of jihad, including the use of economic power to uplift the condition of the needy, and the physical jihad in the struggle against oppression and injustice. The term is also used to denote a war waged in the service of religion. Physical jihad is limited by the following Quranic concepts: "Fight those who fight you, but do not transgress" (2:190); that is, war is justified only if it is defensive in nature. "But if they incline to peace, incline toward it as well, and place your trust in God" (8:61). War is not fought for the purpose of vanquishing or crushing the enemy; peace must be seized at the earliest opportunity. This stress is so important for Muslims that the normal greeting is "Peace be upon you." Islam's peace leaves no room for terrorism, which is the antithesis of jihad. Terrorism is condemned.

A Frustrated Civilization

In general, the West remembers little of the vital role of the Islamic civilization, one of whose centers was in Spain, when Europe was in the so-called Dark Ages. I doubt if the people on the streets of New York, Los Angeles, London, and Paris today are aware of how advanced Islamic civilization was. It provided the world with new knowledge in science, philosophy, literature, law, medicine, and other disciplines. Examples of profound contributions at the turn of the first millennium include those of Ibn Sina, renowned for his work in medicine and known in the West as Avicenna; Ibn Rushd (Averroës) in philosophy and law; Ibn Hayyan (Geber) in chemistry; Ibn al-Haytham (Alhazen) in optics; Omar Khayyam, a renown poet and mathematician; and al-Khwarizmi, known for his profound contribution to algebra (an Arabic word) and whose name is now commemorated in the word *algorithm*. Bernard Lewis described this civilization well when he traced the history of the region: "For many centuries the world of Islam was in the forefront of human civilization and achievement." He adds, "Islam created a civilization, polyethnic, multiracial, international, and one might even say intercontinental. . . . It was the foremost economic power in the world. . . . It had achieved the highest level so far in human history in the arts and sciences of civilization."[4]

I also doubt that people remember that tolerance was a predominant feature of this so-called Eastern civilization. It was during the peak of the Islamic civilization that Muslims, Jews, and Christians lived together peacefully in Spain and other areas of the Muslim world, and it was in the West that the Jews suffered most from discrimination and torture. Cairo was once the place where Maimonides, the Jewish philosopher, studied the ideas of Avicenna and read Aristotle's work, translated into Arabic by, among others, Christian Arab scholars. Using current events in the world today to ignore the contributions of Islamic civilization and to discredit Islam as intolerant is not conducive to world peace and progress.

Unfortunately, some of the problems facing the Muslim world are the making of Muslims themselves. Many in the Muslim world are not aware of the real message of Islam and some leaders and some fanatics use Islam to enhance their own power and political ambition. Moreover, some create new ideologies in the name of Islam and use their interpretations of the Qur'an in debates to drain the human and intellectual power of the society. I doubt if these people truly understand the meaning of enlightenment and the critical role it played in spreading Islamic civilization, not only among Muslims but also throughout the world at large for nearly a millennium. They also may have forgotten that the Qur'an emphasizes the responsibilities of individuals in improving themselves and their societies, stating, "Indeed! God will not change the good condition of the people as long as they do not change their state of goodness themselves" (al Ra'd 13:11).

Today there is a state of discontentment and frustration in the Muslim and Arab world. These feelings are caused by domestic problems and by global or regional political and economic problems. Because of their glorious past, Muslims are asking, *What went wrong?* As evidenced by past achievements, Islam in its proper state is not a source of backwardness and violence. However, one cannot ignore the influence of modern colonization and occupation by Western powers, the disappointment in the alignment with the Eastern or the Western bloc (communism versus capitalism), which failed to yield prosperity, nor can one overlook domestic problems that often result from the ruling by nondemocratic regimes, in many cases supported by Western governments. Moreover, they see through the world media the dominance and prosperity of the West, the humiliation in Palestine, Bosnia, and Chechnya, and their unfavorable economic status in comparison with the rest of the world.

I do not agree with a conspiracy theory of the West against the East; neither do I believe that all the problems are caused by the West. But I do believe that the West should do more to help, as detailed below. Islamic civilization helped Western civilization in the past and it is reasonable to ask for reciprocation now. Furthermore, new methods for better communication are key to continued progress and coexistence. As discontentment and frustrations grow in the have-not world of more than one billion, the world faces increasing risk of conflict and instability, and such troubles will come from boundaries beyond the Arab and Muslim world.

The World of the Have-Nots

In our world, the distribution of wealth is skewed, creating classes within and among populations and regions of the globe.[5] Only 20 percent of the population enjoys the benefit of life in the "developed world," and the gap between the haves and have-nots continues to increase, threatening our stable and peaceful coexistence. According to the World Bank, out of the 6 billion people on Earth, 4.8 billion are living in developing countries; 3 billion live on less than $2 a day; and 1.2 billion live on less than $1 a day, which defines the absolute poverty standard; 1.5 billion people still do not have access to clean water, with concomitant risk of waterborne diseases; and about 2 billion people are still waiting to benefit from the power of the industrial revolution. The annual per capita gross domestic product (GDP) in some Western, developed countries is $35,000, compared with about $1,000 in many developing countries and significantly less in underdeveloped populations. This factor of up to 100 times the difference in living standards ultimately creates dissatisfaction, violence, and racial and ethnic conflict. Evidence of such dissatisfaction already exists; we have only to look at the borders of developed with developing or underdeveloped countries (for example, in America and Europe) or at the borders between the rich and poor within a nation.

Some believe that a new world order can be achieved through globalization to solve such problems such as population explosion, the economic gap, and social disorder. This conclusion is questionable. Globalization, in principle, is a hopeful ideal by which all nations may prosper and advance through participation in the world market. Unfortunately, in its present form, globalization is better tailored to the prospects of the able and the strong, and, although of value to human competition and progress, it serves that fraction of the world's population that is able to exploit the market and the available resources. Moreover, nations have to be ready to enter the gate of globalization, and such entry requires a passage over economic and political barriers.

Barriers to Progress

What is needed to overcome barriers to progress? The answer to this question is not trivial, because many cultural and political considerations are part of the total picture. Nevertheless, I believe that there are essentials for progress that developing and developed countries should seriously consider. For developing countries, there are three essential goals: (1) *building the nation's human resources*, taking into account the necessary elimination of illiteracy, the active participation of women in society, and the need for a reformation of education; (2) *restructuring the national constitution*, which must allow for freedom of thought, minimization of bureaucracy, development of a merit system, and a credible (enforceable) legal code; and (3) *building the science base*.

This last goal is critical for both development and world participation. With a strong scientific base supporting improved education and research, it is possible to enhance the science culture, foster a rational approach, and educate the public about potential developments and benefits. The benefits of science and technology to society are obvious but, just as important, proper science education provides society with rational thinking and thought processes. If absent, a huge void in analytical thinking will be filled with ignorance and even violence. Science is the backbone of progress but, just as important, its knowledge preserves one of the most precious values of humanity—enlightenment.[6,7]

The mindset that such a science base is only for those countries that are already developed is a major obstacle to the have-nots. Moreover, some even believe in a conspiracy theory that the developed world will not help developing countries and that they try to control the flow of knowledge. The former is the "Which came first, the chicken or the egg?" argument, because developed countries were developing before they achieved their current status. Recent success in the world market in developing countries, such as China and India, is the product of their developed educational systems and technological skills in certain sectors—India is fast becoming one of the world leaders in software technology, and products labeled "Made in China" are now all over the globe. As for the conspiracy theory, as stated above, I personally do not give significant weight to it, preferring to believe that nations interact in their mutual interests.

What is needed is acceptance of responsibility in collaboration between developing and developed countries. For the developed world, three essentials are identified:

- *Focusing of aid programs.* Usually, an aid package from developed to developing countries is distributed to many projects (in many cases, most of the aid is for military support). Although some of these projects are badly needed, the number of projects involved and the lack of follow-up (not to mention the presence of corruption) means that the aid does not result in big successes. More direct involvement and focus are needed, especially to help centers of excellence achieve their mission, according to criteria already established in developed countries.

- *Minimization of politics in aid.* The use of an aid program to help specific regimes or groups in the developing world is a big mistake, as history has shown that it is in the best interests of the developed world to help the *people* of the developing countries. Accordingly, an aid program should be visionary in addressing real problems and should provide for long-term investment to ensure true development.

- *Partnership in success.* There are two ways to aid developing countries. Developed nations can either give money intended simply to maintain economic and political stability or they can become partners and provide expertise and a follow-up plan. This serious involvement would be of great help in achieving success in many different sectors. I believe that real success *can* be achieved, provided there exists a sincere desire and a serious commitment to partnerships beneficial to all parties.

Global Returns

What is the return to rich countries for helping poor countries? At the level of the individual, there are religious and philosophical reasons that make the rich give to the poor—morality and self-protection motivate us to help humankind. For countries, mutual aid provides (apart from its altruistic and moral value) insurance for peaceful coexistence and cooperation for preservation of the globe. If we believe that the world is becoming a village because of information technology, then in that village we must provide social security for the less privileged, or we may promote a revolution.

Healthy and sustainable human life requires the participation of all members of the globe. Ozone depletion, for example, is a problem that the developed world cannot handle alone—not only the haves use propellants with chlorofluorocarbons (CFCs). Transmission of diseases, depletion of natural resources, and the greenhouse effect are global issues, and both the haves and the have-nots must address solutions and consequences. Finally, there is the growing world economy. The markets and resources of developing countries are a source of wealth to developed countries, so it is wise to cultivate a harmonious relationship for mutual aid and mutual economic growth.

A powerful example of visionary aid is the Marshall Plan given by the United States to Europe after World War II. Recognizing the mistake made in Europe after World War I, in 1947, the United States decided to help rebuild the damaged infrastructure and to become a partner in Europe's economic (and political) development. Western Europe is stable today and continues to prosper, as does its major trading partner, the United States. The United States spent a mere 2 percent of its GNP on the Marshall Plan from 1948–51. A similar percentage of the $6.6 trillion of the U.S. GNP in 1994 would amount to $130 billion, almost ten times the $15 billion a year currently spent for all nonmilitary foreign aid and more than 280 times the $352 million the United States gave for all overseas population programs in 1991.[8] The commitment and generosity of the Marshall Plan resulted in a spectacular success story. I can see this happening again for Palestine to build a peaceful and prosperous Middle East, and for Africa and Latin America.

It is in the best interests of the developed world to help developing countries sustain a high level of growth to join a new world order and global market. Some developed countries are recognizing the importance of partnerships, especially with neighbors, and attempts are being made to create new ways to support and exchange the know-how; examples include the United States and Mexico and Western and Eastern Europe. The rise of Spain's economic status is in part due to the partnership within Western Europe. By the same token, it is in the best interests of developing countries to address the issues of progress seriously, not through slogans, and with a commitment of both will and resources in order to achieve real progress and to take their places in the developed world.

Building Bridges

Building bridges between cultures and nations is not easy, but the circumstances of the modern world do not permit any culture or nation to remain isolated and insulated. In this century, we are fortunate in having the means to construct such bridges, the mobility to acquire the learning of other cultures, and the human contact that enhances tolerance of other cultures and religions. My own personal experience may be relevant. I am "bicultural." By my fiftieth birthday, I had spent almost equal amounts of time in Egypt and the United States, in the culture of the East and in the culture of the West.

I consider myself fortunate to be enriched by these two cultures, with no culture clash—to gain education in one and contribute to human knowledge in the other, to foster an Eastern tradition in a Western society, and to help facilitate the interaction between the East and the West. This is not new in history. I can envision that the same thing happened when Alexandria, where I received my university education, was a beacon of knowledge. Its famous library, Bibliotheca Alexandrina, brought the West to the East more than two millennia ago.

Science is a universal culture. In the big picture, this universality unites scientists in their search for the truth, no matter what their origin, race, or social background. When I look back at the origins of the science of time and matter, which is central to our research at Caltech, I find a real dialogue. The Eastern, Egyptian civilization I came from was the first to introduce the astronomical calendar around 4240 B.C., measuring accurately the period of a day in a year and, by 1500 B.C., the period of an hour in a day. This was achieved by observing the event of the helical rising of the brilliant star Sothis (or Sirius) and introducing the new technology of sun-clocks or sundials, respectively.

The Western, U.S. civilization I live in gave the world the time resolution of a femtosecond, a millionth of a billionth of a second, the speed needed to record atoms in motion. The concept of the atom, invisible until recently, was given to the world by Democritus of the Greek civilization twenty-five centuries ago. How wonderful and significant that civilizations of different cultures and times have introduced through science enormous benefits to all humanity. It was the rational tradition, in this case of science, that facilitated such building of bridges over millennia of time.

The complexity in world affairs is real and no one can claim that the solutions to world problems are obvious. Whether because of their glorious past or their present geographical and cultural richness, all nations have an important role in helping to solve world problems. As the sole superpower in the world today, the United States has a special role because of its economic, scientific, and military power, but all nations together have responsibilities for a peaceful coexistence in this world.

While the strongest country on Earth must play a fundamental leadership role in combating terrorism together with the international community, it must not lose sight of its leadership role in working for human rights and in reducing the gap between rich and poor, between haves and have-nots. The United States has the opportunity to lead the globe to become a *united* world, to get people all over the world to think of each other as fellow human beings. I vividly remember the American image in the 1960s of a man going to the Moon for the sake of humanity. As Neil Armstrong said in his first words on the Moon: "One small step for man, one giant leap for mankind." The Marshall Plan and the Peace Corps are two examples of visionary initiatives that are representative of that American image of doing great things for humanity.

True, the United States cannot possibly solve every problem in the world, but as the most powerful nation, it should stand tall as a leader and be a role model for others. People around the globe look up to America and many people wish to have an American system of freedom and values. America can be a real partner in helping solve many problems around the world. The reality of the American position was expressed by Zbigniew Brzezinski: "America stands at the center of an interlocking universe, one in which power is exercised through continuous bargaining, dialogue, diffusion, and quest for formal consensus, even though that power originates ultimately from a single source, namely, Washington, D.C."[9]

If history is a coherent and evolutionary process, as argued by Francis Fukuyama, liberal democracy may constitute the end point of humankind's ideological evolution and the final form of human government, and thus it constitutes the "end of history." The argument is supported by the success of the system's economics (free market) and by the successful emergence of the system (democracy) over rival ideologies such as hereditary monarchy, fascism, and communism. This view is controversial, as many believe that Western democracy is not the only viable model of government for the rest of the world; other forms or combinations of systems may be appropriate for different cultures. However, whatever the nature of the system, I believe that human liberty and value, which are basic principles of democracy, are essential for leaps of progress and for the best utilization of human resources. These principles should be exported to the have-nots, but with an understanding of cultural and religious differences, not with hegemony.

Ultimately, with the power of science and technology, and with faith, we will unveil the true nature of our unique consciousness as homo sapiens, the significance of our genetic unity despite race, culture, or religion, and our need for appreciating binding human values. The greatest enemy of human aspiration is ignorance, whether it manifests itself in distorted views of faith, distorted views about other peoples, the failure to recognize the importance and use of new knowledge and new technology, or misunderstandings about nutrition and diseases. It is the source of virtually all human misery.

In this world, we need to build bridges between people, cultures, and nations. Even if we disagree on some issues, these bridges will help us recognize that we live on one globe with common objectives for peaceful coexistence. The key is not to ignore the have-nots, not to ignore the frustrated part of the world. Poverty and hopelessness are sources for terrorism and disruption of world order. Better communications and partnerships will lessen the divide between "us" and "them." We must not allow for the creation of barriers through slogans such as the "clash of civilizations" or the "conflict of religions"—the future is in dialogue, not in conflicts or clashes. We need visionary leaders who make history, not leaders who envision the end of history.

NOTES

1. I wish to acknowledge very useful discussions with Dr. Dema Faham and Dr. Gasser Hathout; Dr. Hathout provided the verses (from the Qur'an) on *jihad*. I also wish to thank Dr. Mary Knight for the careful reading of the manuscript, as well as acknowledge the concise write-up, "Islam: An Introduction," written by the staff of Saudi *Aramco World* (January-February 2002); and, "Glimpses from the Quran," published by the Islamic Center of Southern California (October 2001). Some parts of this chapter are based on presentations I have made in published papers in *Nature* (London), *Proceedings of the Pontifical Academy of Sciences*, and in *Voyage through Time: Walks of Life to the Nobel Prize* (Cairo: The American University in Cairo Press, 2002).

2. Samuel Huntington, *The Clash of Civilizations and the Remaking of World Order* (New York: Simon & Schuster, 1996); idem, "Keynote Address: Colorado College's 125th Anniversary Symposium: Cultures in the 21st Century: Conflicts and Convergences," February 4, 1999; Francis Fukuyama, *The End of History and the Last Man* (New York: Avon Books, 1992).

3. See Karen Armstrong, *A History of God: The 4000-Year Quest of Judaism, Christianity and Islam* (New York: Ballantine Books, 1993).

4. Bernard Lewis, *What Went Wrong? Western Impact and Middle Eastern Response* (New York: Oxford University Press, 2002).

5. Joel E. Cohen, *How Many People Can the Earth Support?* (New York: Norton & Co., 1995).

6. Ahmed Zewail, *The Future of Our World*, in *Einstein — Peace Now! Visions & Ideas*, eds., R. Braun and D. Krieger (Weinehim: Wiley-VCH, 2005, 109); based on the 5th U Thant Distinguished Lecture, United Nations University, Tokyo, 15 April 2003.

7. Ahmed Zewail, *Voyage Through Time: Walks of Life to the Nobel Prize* (Cairo: The American University in Cairo Press, 2002); translated to French, German, Spanish, Chinese, Russian, Korean, and Arabic (appeared in 12 languages and editions).

8. Cohen, *How Many People Can the Earth Support?* Cohen also published "Population Growth and Earth's Human Carrying Capacity," *Science* 269, no. 5222: 341-46; and "How Many People Can Earth Support?" *The Sciences* (Nov./Dec. 1995): 18–23.

9. Zbigniew Brzezinski, *The Grand Chessboard: American Primacy and Its Geostrategic Imperatives* (New York: Basic Books, 1997).

PART III

The Convergence of
the Approaches

৯৯

7

The Convergence of Science and Religion

CHARLES H. TOWNES

An early memorable clash between science and religious ideas occurred in the time of Copernicus and Galileo, who concluded the Earth goes around the Sun instead of the Sun going around the Earth as taught by religions leaders of the time. And as modern science grew over the next several centuries, the split between science and religion became increasingly notable. This included Darwin's introduction of the evolution of species versus a creation event and deterministic Newtonian mechanics that allowed no interference in the course of events by a spiritual force. But things have changed since these early struggles between science and religion.

We currently almost always consider the Earth to be revolving around the Sun but now recognize, from general relativity, that the Earth, Sun, or any other point may equally be considered fixed, and it is equally valid scientifically to consider that the Sun rotates around the Earth. In spite of this change in logic, we still admire Galileo for recognizing the simplicity and value of assuming the Earth revolves around the Sun. With scientific examination of the microscopic world of atoms, scientists discovered quantum mechanics and, with that, Newtonian determinism became philosophically incorrect. Natural phenomena are basically not predictable, though under many circumstances Newtonian mechanics is a remarkably precise approximation and is much used in science and everyday life.

Even after the initiation of quantum mechanics, many scientists continued for some time to believe, in contrast with most religious views, that nature must be deterministic. For example, Einstein postulated a "hidden force" that determined the otherwise apparently unpredictable outcome of quantum mechanical behavior. But recent experimental tests of Bell's theorem show there can be no such hidden force and, in accordance with our present understanding, nature is by no means deterministic. But this does not completely solve the conflict with religious views, because it still seems to say that God cannot actively influence the processes of nature.

It has also been common scientific belief that the universe could have had no beginning but must have always been more or less the same. Einstein, for example, postulated a cosmological force that would stabilize and maintain the universe constant in size or density. The astrophysicist Hoyle argued strenuously against any beginning of our universe and against the big bang initiation of it in particular. But now we know, from scientific discovery of expansion of the universe and from the microwave remnants of its past history, that there was a unique moment in the past, the big bang about 15 billion years ago, when our universe began with near-zero size and expanded to its now huge dimension. Of course, this start of the universe 15 billion years ago, coupled with a well-established age of our Earth of somewhat more than 4 billion years, not only disagreed with the common scientific view of a more-or-less constant universe, it also disagreed with the biblical story of Earth's creation only some thousands of years ago. But how could religious leaders or philosophers of a few thousands years ago understand what the Sun was really like, its relations to the Earth, or the age of the Earth? There was essentially no language or ideas with which to properly discuss the Sun as we now know it, the vastness of our universe, or its origin. And for scientists, it was naturally counterintuitive for the universe to start from essentially nothing.

One can see from these examples that the progress of science has in some cases negated both strong scientific assumptions and strong religious beliefs. But the scientific discovery of a unique moment of initiation of our universe and a nondeterministic future touch on religious beliefs and have some broad parallelism with them. Of course, we cannot expect either scientific or religious knowledge to be perfect or complete, and must be ready for change as we happily understand more.

Science can perhaps be defined as the attempt to understand the nature of our universe, which includes ourselves, and how it works. Religion might be defined as the attempt to understand the purpose and meaning of the universe, including human life. How things work and their purpose or meaning, if there is such, must be inherently related, and a fuller understanding must hence bring science and religion into closer relation and more parallel paths. In my view, if we understand them more fully, they must converge.

Science and religion can appear to be very different—and in a quantitative sense they are quite different. Nevertheless, qualitatively they are rather similar. This is in part because our view of either one depends on our own human characteristics and mechanisms for understanding. These include:

- *Faith or postulates*. It is clear, though generally not noted, that science depends critically on faith as, of course, religion does. For example, we proceed with a faith that the laws of science will not change and are reliable. We make the best postulates we can produce and then proceed with our science, accepting them as long as they appear consistent with experiments or observations.

- *Experiments and observations*. Experiments are usually thought applicable only to science, but observations are also part of the scientific method. An almost purely observational science, for example, is astronomy. In astronomy, we do not modify the things we are observing, as can be done with laboratory experiments; we simply observe. Observation of human behavior, our own emotions, and of history, are important parts of religious understanding, and there can also even be experiments involving religious ideas. An example is the study of the effect of prayer on health, as has been done by Dr. Benson at Harvard and others.[1]

- *Intuition.* This is an important aspect of scientific as well as religious thinking. Most scientists can easily recognize the role played by intuition.

- *Revelation.* From where does a new scientific idea come? It's usually associated with hard work, thinking, and puzzlement over something of intense interest—but just how does a brand-new idea arrive? Is it like Moses' revelation in front of the burning bush after much worry about his people, like the revelation of Gautama the Buddha, sitting under a Bo tree after years of searching for truth, or like Jesus seeing his calling after forty days of thoughtful struggle in the wilderness?

- *Esthetics.* This is perhaps as easily recognizable in science as in religion. Many scientists have noted that a simple and appealing formula or law is most likely right, or "Beauty is truth, and truth beauty," as noted by the poet John Keats.

- *Logic or reason.* Science most obviously and importantly uses logic. But it is also important in religious thinking, allowing conclusions to be drawn from observations of human behavior, of one's own emotions, and encouraging a consistent set of religious conclusions.

I believe that the use of all our human resources or instincts in both science and religion, though with different emphases, produces a qualitative similarity between them, while quantitatively (e.g., in the extensive use of mathematical logic) they can be quite different.

For some time, philosophers have recognized that we cannot absolutely prove or be certain of any of our conclusions. And this has been made very clear in mathematical logic by Gödel, who showed that to apply logic we must start with certain assumptions but we can never prove that these basic assumptions are even self-consistent. Thus, to take our conclusions seriously, we must in the long run rely on such things as intuition, esthetics, and faith. In most cases, our thoughts are also affected by whether our conclusions are acceptable in the judgment of others, our family, friends, and colleagues. And whenever we are able to explore new aspects of our universe, new ideas and conclusions are likely to emerge and revolutionize previous ones. This has happened, for example, as we explored the solar system and the atomic world, and is now happening as we explore cosmology or subatomic phenomena. Many religions have also changed; for example, by giving up literalistic interpretations of instant creation and now finding the big bang initiation of our universe and the long evolutionary development of humans acceptable and qualitatively consistent with previous religious ideas. Thus, scientific discoveries over the last few centuries and even the last few decades have caused many changes in our outlook. I believe some are clearly bringing religious and scientific thought into close interaction, for example, in the recognition of special aspects of our universe.

Characteristically, religion has the view that there is something special about humans, which presumably is associated with God's purpose. Science, however, has commonly taken the point of view that there is nothing special about us or our life. It's all an accidental or random phenomenon—a view that has been strengthened by Darwin's evolutionary ideas, by the multibillion years the Earth has been found to exist, and by the multitude of stars or other suns and planets found in our universe. But there is accumulating scientific evidence that there are indeed special aspects of our universe and our existence.

We do not know why the physical constants have the values they do. But we do now understand that these values are quite critical to the existence of anything like human life. If, for example, the ratio of electrical to nuclear forces were not very close to their actual values, we would not have the richness of chemical elements needed for our existence; the universe might have been almost all hydrogen, or alternatively, all very heavy nuclei. If the properties of nuclear forces and gravity had not been very close to their actual values, there would be no reasonably sized stars of long lifetimes and the energy generation rates we need for the nurturing of life for 4 billion years would not have been possible. Also, if the kinetic energy and mass created in the big bang had not been almost perfectly matched, our universe would have long since been an impossible place for life. And there are many other very specific physical phenomena that we rather recently recognize have come out just right.

Fred Hoyle, a British physicist and cosmologist who has been skeptical of many religious ideas, discovered how carbon is formed by very particular nuclear reactions and wrote in the *Caltech Alumni Journal:*

Would you not say to yourself, "Some supercalculating intellect must have designed the properties of the carbon atom, otherwise the chance of my finding such an atom through the blind forces of nature would be utterly minuscule." Of course you would. . . . A common sense interpretation of the facts suggests that a super intellect has monkeyed with physics, as well as the chemistry and biology, and that there are no blind forces worth speaking about in nature. The numbers one calculates from the facts seem to me so overwhelming as to put this conclusion almost beyond question.[2]

The theoretical physicist Freeman Dyson, at the Institute for Advanced Study, has written:

I conclude from the existence of these accidents of physics and astronomy that the universe is an unexpectedly hospitable place for living creatures. Being a scientist, trained in the habits of thought and language of the twentieth century rather than the eighteenth, I do not claim that the architecture of the universe proves the existence of God. I claim only that the architecture of the universe is consistent with the hypothesis that mind plays an essential role in its functioning.[3]

Thus recent discoveries in the physical sciences, from cosmology to nuclear properties, indicate there is indeed something special about our world that bears on religious thought. They encourage but do not prove the religious view.

To make our life only accidental rather than planned, one can postulate, for example, that there are in fact an enormous number of universes, each with different physical constants and properties, and, of course, humans developed only in the case where characteristics of the universe were exactly right. But this is a very broad and so far untestable postulate, including the supposition that each universe can have different and arbitrary physical constants. As our understanding progresses, perhaps we will discover some logical reason why physical constants of our universe are as they are, or alternatively, why they can have arbitrary values.

The increase in scientific understanding that has occurred over the last couple of centuries is indeed impressive, but it should also help us recognize how much more there is to learn, and that our present basic scientific ideas and philosophy may be overturned as our science increases in depth and understanding. And with further understanding, we must also expect changes in our religious views. The reassuring part of scientific development has been that even with basic philosophical changes, such as elimination of Newtonian determinism by quantum mechanics, our older ideas still have had some validity as a useful approximation. Newtonian mechanics is still extremely useful and adequately complete in many cases. What was tested and believed before, though incompletely understood, still has a certain validity and is of great value. If the pattern of scientific development applies to religious ideas, then as our understanding of religion and human life deepens, there may be both radical changes in our views and at the same time important validity of past experience and ideas. Many religious individuals find, for example, the very distant initiation of our universe and the long process of development of humans a very acceptable and even affirmative modification of earlier religious belief in a more sudden creation.

In science, one can recognize many remaining puzzles and inconsistencies. Physics has many profound ones, such as:

- Why the physical constants are what they are, and why constant— a question already mentioned above. In fact, recent measurements have led to the claim that the fine structure constant, which is based on fundamental physical properties generally assumed to be constant, has been changing over the lifetime of our universe. But this is against the present instincts of most scientists and will certainly be subject to further examination.
- Quantum mechanics and general relativity are both firmly accepted by scientists, but they presently appear inconsistent with each other.
- The creation of our universe, the "big bang," is almost a mystery. We can postulate laws of physics that allow it to happen, but then

how did those laws of physics happen? Any beginning is a mystery. If God created the universe, we are still left with the question of how God came to exist.

- What is the dark matter believed to represent most of the mass of our universe but which we are not able to find or identify?

- Are the zero-point fluctuations of radiation, predicted firmly by quantum mechanics, really present? An experiment suggested by Casimir indicates yes they are, at least to some extent. But the lack of the enormous effect they would have on cosmological observations says they are by no means present as predicted.

- What is the real nature of black holes and their interiors?

- As the universe was formed, why were matter and antimatter created slightly unequally—another queer phenomenon, which was necessary in order for us and the vast galaxies in our universe to exist?

Biology has not recently had to face such apparently basic inconsistencies as has physics, but as biology continues to develop, some may well appear, and there are at least the following mysteries and questions that biology as well as religious thought clearly face:

- Do humans, or other life forms, really have any free will? According to quantum mechanics, our behavior or future is not completely deterministic. But according to present science, there is no free choice. Events on the atomic scale, where quantum mechanical uncertainties or unpredictable randomness occurs, can certainly affect our behavior. But though unpredictable, present science says nothing can determine our behavior in any way except our particular atomic makeup and the phenomena we encounter. Nevertheless, I know no scientist who does not instinctively feel that the individual person can in fact make some choices in his or her actions.

- What is consciousness, or what constitutes a human, and where is it? We all have a sense of what consciousness is. But even defining consciousness seems to defy our present understanding.

- How, really, did intelligent humans develop and how unique are we? It is generally supposed that there are many more intelligent beings living near other stars, and scientists have been searching for signals or other signs of their presence. But quantitative reasoning indicates that if they are as probable as initially expected, then some should probably have visited us. There is increasing consideration that an appropriate solar and planetary system, as well as human-like intelligence, may be extremely rare.

My guess is that as biology proceeds toward both fundamentals and complexity, it may bump into stone walls and radical changes, as has physics over the last century. These changes may revolutionize some biological views while at the same time allowing most present biological ideas to remain useful approximations, paralleling the nature of changes we have experienced in the physical sciences and in many religious views.

In contrast to more or less universal agreement on most scientific issues, there are many differences and disagreements among religions. However, the broad ideas and beliefs of the world's major religions are quite similar. And, as does science, religion faces many major mysteries and puzzles. The growth of science has made some of these religious mysteries ever more challenging, while at the same time modifying and supporting other religious concepts. These puzzles include:

- Can free will exist and what is the nature of consciousness?—problems that face science as well as religion. But religion is still more focused than science on personal responsibility and hence free will, and on the question of when does an individual conscious being begin, where and what is it, and when and if it has an ending. Emergence of new phenomena in complex systems has recently come under active discussion—phenomena we do not envisage when dealing with the simpler units of a complex system, but which can become possible in complex systems such as crystals, or possibly our brains. Certainly emergent phenomena may occur that we have not yet envisioned, and the human brain may provide examples. But whatever phenomena do occur must be consistent with the properties of units from which a complex system is constructed. And according to our present understanding of science, this does not allow us to have free will, a gift we inevitably assume we actually have.

- Why do the innocent suffer? A loving God concerned with humans is a critical part of the major religions. But then why do so many innocent children, and also adults, suffer because of things for which they do not seem responsible?

- How can God act in this world? Quantum mechanics has, as noted, destroyed the idea of completely determined events, but our present scientific understanding allows no room for anything like a spiritual force. Some physicists have postulated that there are additional dimensions to our universe, ones we do not sense. Does this suggest the possibility of a spiritual dimension and phenomena not allowed by the dimensions of space and time we presently understand?

- As does biology, religion faces the interesting question: How unique are earthly humans? We believe there is some substantial difference between us and the animals we know. But are there intelligent beings on other planets? If so, and the question will become most striking if we hear from or encounter them, how do we decide their relation to our religious beliefs? Suppose they are vastly superior in their understanding of our universe and in their religious views, if any. Or suppose they are physically very different and mentally only somewhat more primitive than ourselves? How will such possibilities change or fit our lives and religious ideas? If we clearly recognize the limitations in our understanding of science or of religion, how can we proceed to follow any path with determination? Making the best judgment we can with present understanding, and having faith to follow that judgment, is the only solution.

For some aspects of science, this is easy. Without any second thought we have faith that gravity and mechanics will follow the same laws in the future as today and we can hence plan an airplane trip for tomorrow.

Sometimes scientific faith can be more difficult, however. Consider Einstein's dedication to finding laws that would unify gravity and electromagnetic theory—he had faith in the possibility and worked devotedly at it for the last decades of his life but without any clear success. In both science and religion, we must recognize that our understanding is limited and may change, and also that our faith may in some respects turn out to be unwarranted. Nevertheless, to proceed in science, or if religion is to guide our lives, we must consider what we know or feel carefully, think hard about what is most likely the best path to follow, and then have faith in this judgment that is strong enough to lead us effectively and devotedly through its challenges.

As we proceed, and humans understand more, it seems inevitable that scientific discoveries will make ever clearer the nature of life and our universe and relate ever more directly to questions of meaning and purpose of human life. Already the physics of cosmology, of quantum mechanics, and our understanding of the special nature of our physical world have brought science and religion closer together. I believe we can expect that, with the study of complex biology and the neurosystem, biological science and religious questions are destined to overlap increasingly. We can hope that improved scientific knowledge and insight will very much deepen our understanding, and also that religious thought and understanding will be making parallel progress. Unless arbitrary boundaries are put on what we call the sciences and what we call religion, then as our understanding deepens and broadens, the two can be expected to increasingly interact and overlap.

NOTES

1. H. Benson, J. Dusek, J. Sherwood, P. Lam, C. Bethea, W. Carpenter, S. Levitsky, P. Hill, et al., "Study of the Therapeutic Effects of Intercessory Prayer (STEP) in cardiac bypass patients: A multicenter randomized trial of uncertainty and certainty of receiving intercessory prayer," *American Heart Journal* 151 (April 2006): 934.

2. F. Hoyle, "The Universe: Past and Present Reflections," *Engineering and Science* (November 1981): 8–12.

3. Freeman Dyson, *Disturbing the Universe* (New York: Harper & Row, 1979), 251.

4. S. K. Lamoreaux, *Phys. Rev. Lett.* 83 (October 1999): 3340 or U. Mohideen, and A. Roy, *Phys. Rev. Lett.* 83 (October 1999): 3341.

8

Science and Religion

JEAN KOVALEVSKY

Introduction

The aim of this chapter is to consider the similarities between the rich spiritual and intellectual legacies handed down to us by religion and science. For this purpose, it seems necessary to go beyond conventional representations; to rise above the reductionist visions of an isolated science, on one hand, and of religion as cut off from or overarching the material world, on the other. I would like to present a point of view that, I hope, will contribute to a certain convergence between the very different approaches of science and religion in their quest for a description of reality, a reality that is to my mind unique.

Before tackling this problem, a quick personal remark. Having read only a few of the books on this subject, and since I cannot claim any serious theological or philosophical training, my reflections are strictly personal, and I am acutely aware that the questions I raise will, for the most part, have been dealt with in a far more expert fashion than I could hope to emulate. I draw on my knowledge of the scientific method and thought and the teachings of Christianity. There is no doubt that this attempt to unify these very different approaches to truth is far from new, as we find it in the religions that all have a certain form of cosmology. As far as Christianity is concerned, I will only refer to the Bible, Saint Thomas Aquinas (Thomism is still very much alive in the Catholic tradition), and Teilhard de Chardin.

Antinomies

Science and religion, were there ever two fields more at odds with one another? Let us say that they are antinomical. Admittedly, they have in common that each describes a certain aspect of reality (and perhaps both, in their own way, consider God's creation, even if learned atheists would not admit this). But their paths are quite different and it is hardly surprising that their conclusions are likewise fundamentally different. Science adopts the scientific method, which is a mixture in equal proportions of observations, experiments, and theoretical conjectures. Religion, on the other hand, is based on revelation but also on certain types of experience and historical facts, while exegesis plays an important role in the interpretation of the texts.

We will consider certain aspects that may be less divergent than at first seems to be the case. The difference between each approach and the discourses they give rise to tend to lead either to a situation of conflict or to the annexation of one by the other, or to contemptuous disregard of one by the other.

In the first case, religion is tempted to incorporate within its worldview certain scientific results while rejecting those that seem contradictory. So, for example, whereas some may be tempted to see the big bang as proof of the creation of the world by God, others reject the theory of evolution of species by holding to a literal reading of the biblical account of Genesis. Conversely, scientific materialists, not to mention the proponents of scientism, reject any conception of God and consider that science is self-sufficient. At best, there is a total separation, as science and religion concern two very different worlds. Moreover, this is the position adopted by numerous scientists of faith: for them there are no more points of convergence between science and religion than between music and the building of a hydroelectric dam, and either's existence is not questioned.

Of course, a civil engineer may also be a music lover without composing hymns to the glory of the Colorado dam. But, for my part, I cannot, as a scientist of faith—with a belief in the realities presented by science and religion—dissociate them, and I cannot help thinking that they are two ways of approaching the same *one* thing for which one day we should have a unified explanation. Indeed, the world (and here I am using the word in the sense of everything that exists, as opposed to the word "universe," which refers to the material world as studied by astronomy and the other sciences) is such that each of these two realities can be applied. There should, therefore, be a unifying explanation that takes account of both.

I have already said that science and religion are antonimical. However, this does not mean that they are contradictory. If we look to the *Littré* definition,[1] we can read that it is possible to resolve antinomies. Using Kant's definition of the word (without having to follow his entire logic), antinomy is a natural contradiction that is not the result of faulty reasoning, but of the laws of reason themselves; each time, in a way that goes beyond normal experience, we want to extrapolate something of the absolute from the universe. An antinomy can be resolved with synthesis; and that is our purpose here.

Antinomies in Religion and in Science

If we analyze this notion of antinomy, we find that it is profoundly rooted in Christian doctrine. When an antinomy crops up, it attempts not to produce a choice but rather a synthesis. Hence, for example, "one," "two," and "three" are quite distinct notions. Yet the dogma of the Trinity teaches us that God is a Unity-Trinity (consubstantial and indivisible). In this case, it would be wrong from a Christian point of view to see a contradiction in the form of an opposition between monotheism and polytheism. It is necessary to transcend this apparent contradiction that is the basis of Christian theology. We could also mention the antinomy of portraying Christ as being at once God and man.

This way of resolving antinomies with a synthesis is entirely Christian and can be extended to other couples of opposites such as life and death or body and soul. Hence, death is not a finality marking the disappearance of life but rather a passage that preserves it. The Christian keeps this passage in mind but, in the meantime, he must live life to the full.

The acceptance of the richness of the antinomic duality is not easy. It is possibly even less easy these days, when there is a tendency to oppose and reduce everything to a choice between yes and no. It is the syndrome of all or nothing: the binary 0 or 1 of computers or the logic of the excluded third. For example, to return to the question of religions, Judaism refused the Christ-God antinomy. Later, Islam violently attacked the Trinitarian dogma in the name of monotheism. This is the reason why I think Christian thought is better prepared than that of other faiths to refuse the opposition given by the conjunction "or" and to examine the synthesis provided by the conjunction "and," notably between itself and science. Moreover, in a more general sense, the refusal of such syntheses leads to cut-and-dried position taking, which is a feature of religious fundamentalism and various forms of sectarianism.

Scientific thought, marked by deductive rationalism, notably under the influence of mathematics, also faces significant difficulties regarding antinomical situations. Yet, increasingly, it has to face a number of these situations. Hence, depending on the way we observe it, light can either be described in terms of waves or particles. We know that the acceptance of this antinomy has given rise, since its resolution, to the theory of quanta, one of these best-proved theories of modern science. We can cite other scientific antinomies: Schrödinger's cat that is alive and dead at the same time, the concept of a universe that is at once finite and without limit, the notion that a particle can be in two places at the same time, or the fact that time is a relative concept and advances at a different rate according to the speed at which the clocks are traveling (the Langevin twins), etc.

Parallel to this, Christian thought has divested itself of a literal conformism regarding the sacred texts. The exegeses, based on a better historical and cultural understanding of the Jewish people and their neighbors, and based on forms of deductive reasoning that the purist of the pure scientists could not deny, have progressively unravelled the essence of faith in belief systems and annex traditions. Today, the believer can present his religion in a purer and more rational way than in the past. The church itself, instead of rejecting rational thought, calls for a strengthening of the links between science and faith and not to neglect the contribution of reason to the deepening of faith. Hence, the modes of Christian and scientific thought are converging and have more in common today than they did a century ago. I think we can—we should—go one step further. The truths taught by science and religion are, apparently, antinomic, but, as shown in the preceding examples, we should be able to bring them together in a kind of synthesis.

We have just seen that scientific and religious thought have converged due to the fact that they are able to make a synthesis of certain antinomies. Another point of convergence is the way in which each works toward a deeper understanding and announces its truth where there are also strong analogies. This is what I would like to show now by presenting them in turn.

The Scientific Method

Science is based on observation and experiment. At its disposal it has a certain number of tools such as receptors, laboratory equipment, computers, etc. (I'm simplifying, of course). With the help of these tools, we measure and describe phenomena. But a measure is not simply a number, and an observation is not simply the description of a fact. A measurement needs to be stated in which conditions it was taken (i.e., the temperature, magnetic field, lighting, etc.). In the same way, the facts should be noted down in their context (e.g., does the behavior of an animal indicate fear or aggressiveness? Is it defending its territory?). These details are fundamental, as the next stage is the identification of cause-and-effect relations or of correlations with certain parameters with a view to generalizing the phenomenon by eliminating the secondary conditions. Does the observation or the measurement repeat itself when the experiment is carried out in similar conditions? If not, which parameters need to be modified to ensure repeatability? Which parameters does the phenomenon depend on and in what way? Indeed, two dangers are always looking over the scientist's shoulder: (1) hasty generalizations (all the cats in a town are gray because we saw three in quick succession that were gray); and (2) the invoking of randomness (did the volcano erupt by chance or are there more profound reasons that need to investigated?).

To avoid straying in this way, we rely on theories that describe a certain number of phenomena and which we attempt to use to explain a new one. We sometimes call them "laws of nature" (for example, the law of universal gravitation, the laws of electromagnetism, genetics, quantum mechanics, etc.). These theories remain until new phenomena are discovered that contradicts them. Every new experimental verification adds to the credibility of the theory. However, one experiment carried out properly that contradicts a theory is enough to prove its limitations and to lead to a new theoretical development, and this leads to progress. The new theories include facts that were previously established, as well as new facts. It should be noted at this stage that the scientific imagination is a precious asset; intuition plays an important role in discovery. But it is interesting to discuss the way in which these theories or these laws are presented.

Models

In practice, the expression of these theories or laws takes for granted that "everything happens as if." Newton explicitly employed this phrase when he announced his law of universal gravitation. Later, the observation of the movement of planet Mercury showed that it did not completely match the prediction. So Einstein replaced it with a prediction based on a principle that was entirely different: everything happens as if space is warped by the presence of matter and the trajectories of planets follow appear to be determined by the curvature of space. Despite this new model, which is more accurate, Newton's law remains an excellent approximation. In modern scientific language, the phrase "everything happens as if" is called a "model." This word is revelatory; science does not claim to capture reality but rather provides a description of it or, if we prefer, a transcription of it.

This concept of a model is omnipresent in science, and the development of computers has multiplied their use. We create a model of a star, the climate, a complex molecule, the trajectory of a particle, or the turbulence caused by a plane. We take the physical laws that control the phenomenon, then we write the equations corresponding to these laws and apply them to the object of study in the conditions in which it finds itself. Then we find the solution, which is then compared to the observation, and the hypothesis is modified to satisfy the observations. In this way, we obtain the model of a phenomenon that we can cause to evolve by modifying the parameters.

We can see, therefore, that a model is an abstract construction that allows us to describe an object or a phenomenon, whether it is in the form of analogies, mathematical formulae, a set of hypotheses, graphs, image representations, etc. It is important to emphasize that it is not reality that is reestablished (we don't create a star from a computer simulation!) but a simplified representation in a form that makes it easier to understand or describe. To return to the example of the model of a star, we enter the temperature distributions, the pressures, and the matter found in the interior of stars in such a way that the characteristics observed on its surface (spectrum, temperature, dimensions) are revealed. But nothing proves that the model is the only possible one, and there may well be unknown elements that we have not been able to shed light on until now. Reference has been made elsewhere (in the field of quantum theory and elementary particles) to the notion of "veiled reality," when it is fundamentally impossible to represent a phenomena in its every detail. I dare say that, in the same way, any model only reveals a part of reality in an indirect manner. Scientific truth is, therefore, always presented—and even known—in a cryptic, incomplete, or veiled fashion. To deny, then, that the world is objective, and to consider that everything is images, is a temptation to which some have succumbed, but I won't follow them down this slope that leads to total nihilism.

Another distinctive aspect of science is its power of prediction. The scientism of the nineteenth century, following Laplace, was based on the fact that if we had a perfect understanding of all causes (that is to say, of the laws of nature) and the initial conditions of an evolving phenomenon (for example, the positions of the planets at a given instant), we could infer with absolute accuracy its evolution in the future. We now know that certain laws of macroscopic physics have a statistical quality based on the laws of large numbers (the second principle of thermodynamics), whereas, in particle physics, there is a fundamental uncertainty (according to Heisenberg, we cannot observe to any great degree of accuracy the position and the speed of a particle). Other perfectly deterministic laws, like the law of universal gravitation, can lead to situations of instability, leading to an uncertainty regarding the evolution of a system to the extent that we cannot know with infinite precision the conditions at any given instant (deterministic chaos). We have also introduced the notion of quantum chaos. All this adds to the elusiveness of physical reality and to the limitations of the models that attempt to describe it.

To complete this overview of the representation of scientific reality, we need to point out another danger from which the image of science sometimes suffers. It is certainly good to present to the public the results of scientific theories, but often the process of vulgarization oversimplifies the models. As a result, this gives a simplistic, unfortunately often erroneous, image of reality.

Religion

The particularity of religions is that they are based on a revelation, but this is not enough. It is not enough to declare oneself a messiah or a guru to impose the message we've received (or that we think we've received). Isolated confirmations are not credible enough to establish a religion. Revelation is only really accepted as such when it is accompanied or is nurtured by numerous personal or collective, lone or shared, experiences. It may be a case of observed facts, testimonies, mystical or spiritual experiences, meetings, thoughts, sudden or progressive conversions. Some of these experiences are esoteric; others, transmittable. For example, it is this transmission that underpins the strength and continuity of monastic orders. It is the accumulation of these events that constitutes the soil in which religion takes root, comforted by tradition and doctrinal developments and that, all together, are just as impressive as the basis of a scientific theory.

Insofar as the description of religious facts does not require the same rigor as scientific measurements or observations, and that they tend to be the result of personal or emotional interpretations, they are better adapted to cultural and philosophical environments. Yet science is riddled with errors linked to prejudices. To give a recent example, the theories of Lyssenko in the 1950s show that science is not always sheltered from errors associated with preestablished political or philosophical a priori.

Let us consider the example of Christianity. Its fundamental revelation is found in the Gospels, even if it was heralded in the revelations of the Old Testament. The Gospels relate facts and communicate the teaching of Christ, and this contributes to the establishment of historical veracity and the contents of dogma. I would like to illustrate this point by drawing the attention of the reader to what we are told regarding the kingdom of God. It is related in the form of parables. But what is a parable if not a profound and indescribable truth conveyed by analogy, with the use of an image or a narrative that is suggested by the listener's environment? Hence, the kingdom of God is presented by several parables that start with the words "to what will I compare the Kingdom of the Heavens? It is similar to. . . . " It is precisely the same as a scientific model. The love of God for humanity is likened to the love of a father for his prodigal son, or to the company boss who gives a full salary to his workmen even though they have only worked for an hour. Again, they are models. I would even say they are forms of vulgarization.

The Orthodox Church also presents models to its faithful in the form of icons. At first sight, they are stylized representations of individuals or events, quite different from the religious paintings of the West. They are models that the believer, through his or her veneration, interprets like windows onto the kingdom of God, and this contributes to the strengthening of his or her faith. It is through these icons that he or she comes into contact with this ineffable religious reality. However, receptivity to these representations is also, partly, culturally biased. Others may be more receptive to different models or symbols like the Easter candle or the sites where the Virgin Mary has appeared.

In religion, these symbols and models are even more distant from the reality they hearken to than in science. Moreover, religious knowledge, to a much greater extent than scientific knowledge, is partial and imperfect, and its transmission is certainly more simplistic and distorting. We can say, therefore, that religious reality comes to us in the same way that scientific reality does, in a veiled fashion. The additional difficulty is that the interpretation is more personalized, which can explain the diversity of the main branches of the Christian tradition.

Interpenetration of Science and Religion

The proverb *à chacun son métier et les vaches seront bien gardées* (each to his own job and the cows will be well looked after) describes the position that is often adopted: science should have the task of unveiling "how" natural phenomena occur, while philosophy and religion find out "why." It's clear but simplistic, since the two are interested in the same world. Is not the best way to avoid conflicts to encroach on the other side's territory? This amounts to refusing the other's approach, and to reduce one to fundamentalist dogmatism and the other to sectarian materialism or scientism. These two extreme points of view are far too rigid. It is tantamount to resolving the antinomy between these two approaches by a permanent binary separation. Yet a dialogue is necessary in view of the need to provide a synthetic answer to certain fundamental questions that both have in common. Equally, it is not possible to answer the question why without knowing how. By the same token, a global vision of the world cannot be dispensed with the explicit or implicit philosophical interrogations of the big problems related to the soul. Let me give a few examples.

The universe made it possible for living then thinking creatures to emerge. Before discussing why this was possible, physics needs to frame the question properly. It is up to scientists to tell us about the necessary fine-tuning of the universal constants making it possible for heavy elements to form in the heart of stars, for the universe not to implode before the apparition of life and for complex chemical reactions to be produced under certain conditions, and for the basic constituents of biology, once formed, to remain stable, etc. If, as certain calculations seem to suggest, the margins of probability are extremely narrow, then the question of chance should be addressed in science and in religion—and it would be dishonest of either one to discount it out of hand, even if we can expect that several answers will be proposed.

Another example is given by the observation that, contrary to the predictions of statistical mechanics regarding thermodynamics, we observe a frequent (if not a general) tendency toward the formation of increasingly complex elements (heavy atoms, simple molecules, and those that characterize life). We can see that this tendency leads to progress, and this immediately raises the question: Why? Science and religion both have something to say (this is what Teilhard de Chardin attempted to do).[2] The one without the other, however, will only provide a partial answer to the question. Scientists will tend to invoke a positivist or metaphysical precondition, whereas a strictly religious interpretation, not based on scientific results, leads to the crudest form of creationism.

We could even inquire further into the mysteries of life—into the nature or origin of life—by contrasting the religious approaches to these problems with the advancements of science. The same could be said of the origin of the universe or of the destiny of humanity. Although it is always veiled, what will come out of this synthesis should be able to claim greater legitimacy in our approach to deep reality. It is, in any case, in this direction that we need to go to resolve the antinomy between science and religion, these two classes of approach to truth. To go further, it is useful to consider this notion of the truth of the points of view of science and religion.

Scientific Truth and Religious Truth

In their own separate but not, for that matter, mutually exclusive ways, science and religion are both involved in the search for truth and how to communicate it. Despite their obvious differences in approach, they nonetheless partly derive from the same logic. We have seen that the results of scientific research are presented in the form of models that we then try to interpret with our observations and the fundamentals of physics. These models allow scientists to make intelligible representations of observations and measurements and to account for the repeatability of the effects when the causes are fixed.

In the case of religion, in which dogmas play an analogous role to the laws of physics in science, the models are continuously challenged by the religious and spiritual experiences of believers. The coherence of the whole is the objective of theologians. I would like to insist on this analogy. In the same way that scientific theories evolve when observation requires them to, theology can also be enriched when there is a consensus among the faithful. Exegetes and theologians are to religion as theoreticians are to science. In the same way that major scientific experiments and observations lead to laws that are recognized by the entire scientific community—and become essential for our modeling of material reality—mystical experiences are recognized by the adoption of new dogmas by Catholics, a more gradual and subtle progression by the Orthodox Church, and a certain liberation of concepts by Protestants, not to mention by beatifications or canonizations, by the creation of new religious orders, or simply by a consensus, and contribute in this way to enriching religious truth and its tradition.

This could be because the two approaches are reflections of the way the human spirit operates, but the fact is that the search for truth follows similar and equally rigorous processes in science and religion. Of course, the "proofs" of these truths are not of the same order, although they are the result of similar undertakings. The proof of constrained relativity is found for example in the measurements that are taken in particle accelerators. Proof of the existence of God can be found in the accounts of mystical experiences. Of course, not everyone can personally renew their mystical experiences. But is this a reason to deny them or to deny their significance? I defy the man on the street to try and repeat a particle acceleration experiment. But is this a reason to deny the results obtained and their theoretical consequences? Not everyone is Einstein or Saint Teresa of Lisieux. Hence, the problem is one of communication and adherence to these truths.

Trust and Faith

I am a scientist. I understand my colleagues' research methods and procedures, but 99 percent of the time I am powerless to verify their experiments or the conclusions or models they present. To me, using the example of very diverse sciences—the action of the RNA or of neurotransmitters, the calculating of predicates, quark theory of the organization of liquid crystals—is double Dutch. Yet I trust my colleagues and believe in their results, just as they have trust in me when I present my scientific specialty that, generally speaking, they do not understand any better. I trust them even if they have recourse to several different models in the search for explanations. Indeed, there is a whole web of connections between the different sciences that encapsulate all scientific findings in a coherent whole.

This does not preclude the need for a critical spirit; there are scientific errors (and even false scientists). We hear it said, and I have said it myself, "I don't believe this result." This happens when it concerns an area I know well or because it contradicts my own experience or because it goes against my personal views (views that are necessarily limited) concerning the physical universe. If the results are confirmed, however, I must accept them and change my ideas. But it also happens that results that have been announced as having been seriously checked, often highly publicized, in the end turn out to be imprecise, thus justifying the skepticism with which I received them (for example, the memory of water, the fifth force, or sniffer planes). These heterodox results have an outward scientific allure, and it is essential to spot them as quickly as possible because they can do considerable harm to science. Hence, in spite of these errings, but with confidence in the community of scientists to separate the wheat from the chaff, I believe in the results of science.

I am a Christian. I appreciate that there can be a religious view of the world. Moreover, I deeply believe that such a view is necessary. However, I am 99 percent incapable of verifying what I am taught. I have not had any mystical experiences, I am a hopeless theologian, and I am incapable of making conversation on original sin or the Immaculate Conception. However, I trust the many individuals who have testified about their mystical practices, and those who live their faith at home or in monasteries. I also have faith in this bi-millennial tradition that is the product of the immense accumulation of experience and that constitutes this coherent and harmonious whole. In other words, I trust all these accounts, I trust their implications, and I dare say this trust is growing in me. In this area, mistakes are also made. It has been said that a religion is a sect that has succeeded; just as in science, there are many ideas that have not succeeded. If sects do not succeed, it is because they are not backed up with enough similar testimonies upon which to build a religion. In the same way, science is the result of a process of selection.

This parallelism leads me to place the same trust in Christianity as I do in science. This is why, in these conditions, it seems indispensable to me that a vision of the world should include elements that are drawn from each area in a mutually enriching synthesis. Achieving this, however, will require resolving their antinomy by means of a confident dialogue. These two visions should complement each other and even converge to form a much more powerful tool for the advancement of knowledge and our understanding of the world. Personally, such a synthesis allows me to reconcile my Christian faith with my life as a scientist. Ultimately, such a synthetic vision allows me to find meaning in the world, without which I would have a sense of incompleteness.

There is one unassailable fact that we cannot avoid: despite a few variations, scientific understanding is unique, whereas there are several large religions. This allows materialists to state that this proves that religions are devoid of meaning. However, when looked at in more detail, it is clear that there is a common thread running through all religions: the force of mysticism; the strength of the spirit (that can interact with the material world); the acceptance of God (whether he be one or take several forms) and a certain capacity to communicate with him or them;

the not unavoidable or definitive characteristic of death, etc. It is certain
that the followers of other religions use the same reasoning as I do to
justify their faith with just as much conviction. If we compare the objective of a religious life to the climbing of a mountain (another model!), it
is quite possible that there are several ways to the top. But it is essential
to follow a path we have chosen and that we trust, as trying to change
path halfway along would be to run the risk of getting lost or falling
into a precipice. This is why I am not bothered by the fact that there are
other religions that are highly respectable and, possibly, just as effective
for reaching the summit. Each one can add something to the purely scientific vision of the world.

The history of science gives us a point of comparison. Indeed, it has
appeared in very different countries. Chinese, Assyrian, Inca, Egyptian,
and Greek sciences were all very different. Yet they all produced valid
descriptions and models in the areas of astronomy, geometry, metallur-
gy, and mechanics. In the same way, sciences belonging to a common
field have considerably changed over the centuries. The sciences of the
Middle Ages or from the seventeenth or nineteenth centuries are quite
different from modern science. Yet all contain a part of scientific truth
that is more or less approximate, more or less well explained but correct. Who can say what the science of the twenty-second century will be
like? Why should that which applies to science not apply to religions if
they have a certain basis of truth, which is more or less big, more or less
well expressed?

Example of Synthesis

We come, then, to the following reasoning: There is a scientific truth and
a religious truth. Moreover, by definition, the world is unique, since it
contains all that exists. Yet God exists (it is the first truth of religions);
consequently, he must be in the world. Therefore, a complete cosmology
should include him. Besides, if the scientific study of the material universe, with its three dimensions plus time, has not made it possible to find
God, then the conclusion must be that he transcends the material universe

and the world that concerns us should be extended to other dimensions.

Just as a three-dimensional being could touch a hypothetical two-dimensional being without the latter being able to conceive what he was like, God could be among us and even within us without our in the least bit suspecting it. In such a hyperuniverse, which is not easily accessible except with the help of models but which mathematicians can define, God and the material universe either coexist or they form a symbiosis. Such a possibility would make it possible to solve the science-religion antinomy while explaining the impossibility of representing God. Qualifiers such as "love" or "life" that we attempt to attribute to him are not measurable but, according to the Christian religion, allow a fragment to be present in us ("the kingdom of God is within us"). Hence, man would have an element of himself in this other dimension, which is exterior to our material universe. We could call this element "soul."

In such a vision, religion and science play a complementary role. Today certain properties of the wave-particle association (Aspect experiment) are not comprehensible by science, and it has been suggested that another dimension be considered to explain the instant transmission of information. This is not the divine dimension, but this only goes to show that to limit oneself to that which is directly accessible—to physical observations and our current models—is insufficient.

There are also other areas that are taboo in modern science but that could shine a glimmer of light on the relations with the spiritual universe. I will only cite as examples hypnosis or certain premonitions, not to mention the visions that are catalogued in religion. It is a slippery terrain onto which one should only venture with the utmost caution. Nonetheless, I think that to enquire into these areas would allow science not so much to meet God but to observe that the substrate in which religious truth lies has an objective existence.

Conclusion

As a conclusion, I would like to reproduce the following paragraph from *Le phénomène humain*:

The time has come to realize that an interpretation, even if it is a positivist interpretation, to be satisfactory, should cover the outside as well as the inside of things, the spirit as well as matter. Real physics is the physics that will one day succeed in integrating Man into a coherent representation of the world.[3]

NOTES

1. One of the standard dictionaries for the French language.
2. Pierre Teilhard de Chardin, *Le phénomène humain* (Paris: Éditions du Seuil, 1955).
3. Ibid.

9

Moral Philosophy

A Space for Dialogue between Science and Theology

THIERRY MAGNIN

Introduction

It is now generally accepted that the development of the hard sciences in the twentieth century (in particular in the areas of mathematics and physics) has led to a reappraisal of the traditional philosophical notions of reality and meaning. In the area of the epistemology of science, the emergence of a new vision of complexity in the fields of quantum physics, thermodynamics, and cosmology has resulted in a redefinition of the word *reality*, perceived in scientific research as the relationship between subject and object.

The observer is part of the reality that he analyzes. The theory of measurement even demonstrates that, for a physicist, to know and to measure is to act on reality. It is a "reality of interactions" that is put to scientific analysis leading to a profound reevaluation of the three dogmas of scientism: Laplacian determinism (notions of unpredictability and uncertainty), ontological reductionism (there is more information in the whole than the sum of its parts), and methodological reductionism (undecidability and incompleteness with the Gödel theorem). As a result, man has become an "interpreter" of a complex world, throwing into contention such notions as the strong objectivity of the sciences.

In light of this progress, questions such as "What is at stake today?" and "What choices and values are implicit in the scientific method?" have been asked. Einstein is particularly interesting in this regard: "Let us concede that behind any major scientific work is a conviction akin to religious belief, that the world is intelligible!"[1] The very term "conviction" is indicative of scientists' preestablished choices and moral attitudes.

The objective of this chapter is to analyze these attitudes (the field of the ethics of knowledge), to determine where they come from (moral philosophy with critical analysis of these belief systems), and to have a physicist's view on the topic of science and conscience today. For this, we will need to clearly distinguish between the fields of science, metaphysics, and moral philosophy.

Scientific Reasoning within the "Game of Possibilities"

As an introduction to some of the defining elements of current scientific method, I quote the concluding lines of *The Game of Possibilities* by Francois Jacob, Nobel Prize–winner for medicine.[2] In this book, Jacob is particularly interested in the relationship between the *hard* sciences and myths. He begins by stating that science can be credited with having largely contributed to dispelling the idea of an intangible truth. He then goes on to demonstrate how numerous human activities (the arts, the sciences, technical education, politics) are just different ways of "playing within the game of possibilities," each with its own set of rules. In many respects, he considers that science and myths have similar functions. They both provide the human spirit with a certain representation of the world and its dynamic forces that sustain it. They both define the range of possibilities.

Whether it be in mythical or scientific way, the way man views the world is largely conditioned by his imagination. In Science, to obtain valuable observations it is necessary to have some prior idea of what there is to be observed. It is necessary to decide beforehand what is possible. This approach is obviously guided by an idea of what reality is. It always involves a certain idea of the unknown, of that zone just beyond the boundaries of what logic and experience allow us to believe. Scientific research always begins with the invention of a possible world or a fragment of a possible world. This is also how mythical thought originates, though the similarity stops there.[3]

Having outlined the similarities between science and myth and the fundamental role of imagination in both cases, the author goes on to highlight the basic differences between these two approaches of what is possible.

What should be noted in Jacob's analysis is the way in which man seeks to "invent the future" through different activities, each with their own set of rules, but which all rely on his imagination. His conclusion leads us to an ethical consideration.

Our imagination projects the constantly renewed image of what is possible, and it is against the background of this image that we are forever confronting our fears and hopes. It is against this representation of the possible that we check what we love and what we loathe. However, if it is in our own nature to work toward the future, the system is such that our forecasts must remain uncertain. We cannot imagine ourselves without a future, but we cannot imagine what that future holds. Changes will come about; the future will be different from what we imagine. This is particularly true of science. Research is a never-ending process; we can never see where it will lead. Unpredictability is an inherent part of scientific activity. It is necessary to accept its unexpected and alarming aspects.[4]

Most leading scientists subscribe to this view and some, such as Poincaré, Hadamard, Heisenberg, or the late Bohr have had the courage to declare so publicly.

When Science Is Confronted with the Complexity of Reality

Since physicists have been confronted with the complexity of reality, scientific thinking as a whole has undergone a profound upheaval as evidenced by the demise of the Laplacien dream, "the end of certainty," or the withdrawal of foundation.

The Demise of the Laplacien Dream

Classical science was dominated by the notions of permanence and stability, predictions, determinism, and, ultimately, control. The idea of certainty in science was widely held and virtually synonymous with the "sharing of divine science." The development of quantum physics and non-equilibrium thermodynamics introduced the concepts of uncertainty, incompleteness, and undecidability into the sphere of rationality, which radically alter the status of knowledge and the place of the knowing subject. There was a radical change in scientific thought that had a marked effect on people's way of thinking in general.

Poincaré and many later scientists demonstrated that the "Laplacian dream" of determinism was an illusion. If, indeed, Newton's laws allow us to accurately predict the movements of two bodies in motion insofar as their precise trajectories are known, the same cannot be said of systems comprising three or more bodies. Complete predictability is impossible; there is no general solution to the problem. Poincaré is also responsible for the notion of unpredictability that characterizes deterministic chaos (the unpredictable behavior of a system despite the fact that it is described in terms of the equations of determinist evolution). This determinist chaos can often be seen in nature. Sensitivity to the initial conditions renders the Laplacian dream obsolete: it is not because a system is subject to a formal determinist evolutionary law that this evolution is predictable. Therefore, in respect of our current understanding, a complete description of reality cannot be conceived of.

It is important here to emphasize a key point. By replacing Laplacian determinism and the idea of certainty with determinist chaos and unpredictability, scientists opened new avenues for scientific progress. The idea of certainty seemed to be the only worthy basis on which to build a genuine scientific enterprise. However, this vision was, as it turned out, pessimistic, and time (and its arrow) was just an illusion.[5] Unpredictability and chaos restored the role of time, allowing it to play a constructive part in an "uncertain reality."[6] Here, the idea of probability was not introduced as a result of our ignorance but as the very result of evolution! Non-equilibrium gave us an idea of the potentialities of matter. Needless to say, then, this change in worldview was bound to have an effect on the attitude of the scientist. For us scientists, the universe is not a given—rather, it is under construction!

Something of Reality Is Beyond Our Knowledge

Science's claims of "completeness" that go hand in hand with its claims of certainty presuppose the existence of a language that reflects the totality of reality. Wittgenstein's studies demonstrate that the logical structures of language cannot be written within language itself.[7] In other words, the medium in which (or thanks to which) we represent things is not representable (it cannot be expressed). There are concepts that are inexpressible outside of language. Is not the acceptance of the inexpressible a way of opening the door to the question of meaning while recognizing the contingence of man?

Classical science, with its dream of perfect predictability, acknowledged its ambition to construct a comprehensive system of representation. But Gödel's work put an end to this. His findings clearly demonstrated that there are undecidable propositions, true arithmetical propositions that cannot be deduced from axioms, and truly irrefutable statements.[8] Consequently, no theory can, of its own accord, provide proof of its own consistency; complete self-description is logically impossible. Consistency, therefore, implies incompleteness, and completeness can only be obtained at the expense of consistency. Here again, what progress!

Quantum physics is prime ground for showing incompleteness, this "thing that is beyond our knowledge." Microphysics reminds us that man is not an independent spectator of the reality he explores but an integral part of it (we are "of the world," *in situ*). The reality described by physics is no longer independent from the terms of description. This is not only because, as we know, man developed these concepts and theories, but also because to measure and to know is to have an effect on reality or, rather, to interact with it. This interaction by definition modifies the object. Consequently, each measurement is marked by an irreducible indetermination expressed, in quantum mechanical terms, by Heisenberg's uncertainty principle. This uncertainty appears, then, to be coextensive of the knowledge we derive from reality. There is a real limit to our knowledge of the quantum object. Something eludes us yet knowledge also progresses through the nonpassive acceptance of this incompleteness. I emphasize nonpassive acceptance, as Einstein's resolve to find the flaws in quantum theory (the search for hidden variables) was a contribution to the progress of knowledge.

Something is beyond our knowledge, something of the order of origins. Whether it be in the study of language (Wittgenstein), logic (Gödel), the structure of matter (Heisenberg), or irreversible evolution (Prigogine), it is apparent that similar conclusions are being arrived at regarding incompleteness, the horizon of undecidability, and the impossibility of limiting truth to the totality of what can be said, whether this be formally demonstrated or directly measured. To accept that something can be formalized is to accept that some aspect of that thing is necessarily missing. Constructing a theory of knowledge requires us to accept that something is beyond our knowledge. This does not represent a defeat of reason. Rather, it is a condition of progress, of intelligibility.

The classical concepts of linear causality, reduction, completeness, and stability are replaced by those of sensitivity to the initial conditions, irreducibility, incompleteness, uncertainty, instability, and unpredictability. Moreover, contemporary science invites us to get a measure of the positivity of this incompleteness that even appears to be a condition of knowledge. It is a good introduction to the question of the significance and the place of the subject in the exploration of the world of which it is a part! This is how scientific knowledge has progressed from certainty to uncertainty—and we are reminded of the contingence and finite nature of man.

The Withdrawal of Foundation

A characteristic of epistemological thinking today is to note what Ladrière has called "the questioning of the foundation, indeed, the withdrawal of the foundation."[9] According to Ladrière, this observation can be made through Hilbert's project to found mathematics on logical atomism and the development of phenomenology (the attempt to re-constitute the movement of the self-construction of experience).

In these three cases, the method consists in discovering a privileged region that contains the guarantees of its own validity and showing how, through appropriate actions, it is possible to shed light on the relatively obscure parts of the discourse on experience without prejudice to the region (the role of foundation played by this region). Ladrière shows that Hilbert's project on the foundation of mathematics has come up against the limits of formal systems:

The demonstrations of non-contradiction (which are the main components of Hilbert's program) can only in part be relative. The idea of a privileged found-ing domain is untenable (both because there is no way of "reducing" every-thing to such a domain and because it is impossible to identify a region which would be capable of founding itself in an absolute sense).[10]

According to Ladrière, what serves as a foundation at any given moment only constitutes a temporary pause in a process that is bound to contin-ue. These are only the contingent conditions of the research, the tem-porary limitations of operational, conceptual, or experimental means of investigation.

There is, therefore, no essential difference between the founder (who is only ever improperly the founder) and the founded. There is no true discontinuity in their status. This signifies that this type of unshakeable solidity, this faultless consistency which was attributed to the foundation and which was transmit-ted to all that was founded is no longer looking so sound.[11]

Bohr's Complementarity Principle and Its Confrontation with Complexity

The history of Bohr's idea of complementarity has been examined by Gérard Holton.[12] The key points of his argument are complementarity in quantum mechanics and the question of different levels of reality.

Complementarity in Quantum Mechanics

In quantum mechanics, the description of elementary particles (like the electron) that make up matter requires the use of terms that appear to be mutually exclusive, which we will call "contradictory" or "antagonistic" (*A* and non-*A*). For example, an electron is a well-recognized elementary particle whose trace and impact can be picked up by a detector (corpuscular properties). But its wave properties are just as well established and are exhibited in the phenomena of diffraction (with interferometry). To describe a particle, quantum physics refers to wave *and* corpuscle, even if experimentally the wave characteristic *or* the corpuscular characteristics are exhibited independently.

These two images of wave and corpuscle are mutually exclusive. In fact, a given entity cannot, at the same time, in our accepted usage of language, be a wave (that is to say, a space that extends to a greater space) and a particle (that is, a substance enclosed in a very small volume). With complementarity, however, continuity (the wave aspect) and discontinuity (corpuscular aspect) will be considered at the same time in the description of elementary particles. In this way, we find that there are numerous examples of contradictory couples (or antagonisms) in quantum mechanics: continuity-discontinuity, separability-nonseparablity, symmetry and broken symmetry, local causality and global causality, for example. Thus, a system composed of two elementary particles that is said to be entangled (both emitted by a same source, for example) is said to be *nonseparable*. Nonetheless, the logic we derive from everyday life indicates that our macroscopic world is made up of separable elements even if interactions between these elements exist and can be determined. The question is, then, how to reconcile continuity and discontinuity, macroscopic locality and microscopic locality?

Among the different approaches proposed for resolving this question, the most convincing is the principle of complementarity as expounded by the physicists Bohr and Heisenberg. They believed that complementarity describes a phenomenon by two different modes that are necessarily exclusive. It is only by considering these two contradictory modes that one can start to understand the phenomenon: "When playing with these two images (wave/corpuscle for example), going from one to the other and then back again, we finally obtain the right impression of the strange sort of reality which hides behind our atomic experiments."[13] Bohr and Heisenberg made use of the concept of complementarity on several occasions in order to interpret quantum theory. Hence, knowing the position of a particle is complementary to knowing its movement quantity (product of mass times velocity). If we know the value of one with a high degree of accuracy, then we cannot know the value of the other with the same degree of accuracy (Heisenberg's uncertainty principle). Yet we need to know both in order to determine the behavior of this particle.

A particle can be studied experimentally with a detector or an interferometer. In other words, according to Bohr, if one wants to talk about a quantum object, it is better to do so in terms of corpuscle or wave, depending on the way the experiment is set up and in relation to the question asked by the observer. No image is ever complete and it is necessary to make use of two contradictory images to describe the quantum object. The change this represents, compared to classical physics, is that the very definition of the physical measurements is directly affected by the procedures and measures used: "The measuring procedure has a fundamental influence on the conditions on which the very definition of physical quantities in question is based."[14]

In this way, Bohr was able to show how in quantum mechanics the fundamental premise of the indivisibility of quantum action forces us to adopt a new method of description that can be called complementary. Any given application of classical concepts prevents the simultaneous use of other classical concepts that in a different context are equally necessary for the elucidation of phenomena. Let us emphasize here the importance of the coupling of *experimental conditions and the conceptual apparatus* that forms the basis of Bohr's complementarity principle. This principle is intended to determine the manner in which those concepts work, which plays a part in the understanding of the theories of quantum phenomena—a fundamental concept for the philosophical analysis of the idea of complementarity.

This "way of viewing reality" gives rise to a paradox, at the level of language, as in the case of the wave/corpuscle, locality/nonlocality couples. However, for Bohr, the paradoxes resulting from these double descriptions are, so to speak, put to one side by the fact that it is impossible to take two simultaneous measurements of the same object, those of its wave characteristics and its corpuscular characteristics. When one of these images is materialized, the other becomes virtually or potentially realized. Let us stress, however, that this complementarity has more to do with mutually exclusive aspects of quantum phenomena than a mere juxtaposition of images. The elementary particle is neither a wave nor a corpuscle but a "thing" that combines the two images.

Different Levels of Reality?

The philosopher and scientist Stéphane Lupasco and the physicist B. Nicolescu made two little-known but nonetheless major contributions to the idea of complementarity.[15] Lupasco's general idea was to propose a new logic based on what the experience of microphysics was able to say and reveal about human thought. According to him, although Hegel and Bachelard were aware of the fact that classical science was ill-suited to describe microphysical experiments, they did not go far enough. Refuting classical yes-or-no logic, Lupasco showed that only the logic of the included middle is capable of taking into account complete reality. The diversity of reality can be structured and contained in the triad, Actualization (A)—Potentialization (P)—State (T) (which corresponds to the included third term). The actualization corresponds to that which is experimentally measured. Potentialization is that which exists "potentially" even if it is not actualized (for example, the physical states corresponding with the wave function). State T implies a dynamic equilibrium between A and P. Basarab Nicolescu introduced the concept of levels of reality into Lupasco's system.[16] To properly understand this concept and in order to avoid confusion with closely related concepts of *levels of representation* and levels of organization, we offer the following analysis.

When the physicist wants to describe a quark, for example, he starts by describing it as a purely mathematical entity (this is the first level of representation), then as a free particle (the second level), and, more recently, as a particle confined in the hadrons (the third level). In fact, these three levels of representation belong to the same level of reality, which we shall call the quantum level. Conversely, quantons (which correspond to a particular level of representation of the elementary particles) also correspond, as we have seen, to waves and corpuscles (another level of representation). But, in this case, these two levels of representation correspond to *two levels of reality*, to the quantum and classical levels in physics. At the level of the organization of matter, representations are either at the same level of reality or a combination of several levels. Thus a level of reality will correspond to a family of systems that remain invariant under the action of one law. One can distinguish different levels according to the scales used: at the level of particles, man, or planets. Moreover, two levels of reality are different if there is a break in the laws, the logic, or the fundamental concepts (like causality, for example) when one passes from one level to another.

At a macroscopic level (level 1), the local causality (and separability) is dominant, but at the microscopic level (level 2), causality is global and there is a non-separability.

What appears to be contradictory at level 1 (wave-corpuscle, separability-nonseparability) can be united at level 2 with state T linked to the dynamic of the antagonists.

Without an appropriate transition in the passage of one level of reality to another, any number of paradoxes can be created. By this approach, the logic of the included middle is not at odds with the principle of noncontradiction thanks to the concept of levels of reality. It does, however, highlight the need to take the included middle into account.

This summarizes the idea of complementarity and its development.

It is also worth pointing out Edgar Morin's work on complexity and the idea of complementarity,[17] in the same vein as Nicolescu reported here but without the enlightening concept of *levels of reality*. This idea constitutes a particularly fitting example to highlight the evolution of scientific mentality in respect to the complexity of reality as it is analyzed.

Moral Philosophy: Common Ground to Be Explored

According to the leitmotif of modern epistemology, in the analysis of the incompleteness of science as a whole something is beyond our knowledge. The principle of complementarity is an interesting illustration of this. Contemporary science invites us to measure the positivity of this incompleteness, which now appears to be a very condition of knowledge. It is a good precursor to the question of the meaning and place of the subject in the exploration of the world to which it belongs. There is a withdrawal of foundation; "something is beyond our knowledge." This "absence of fixed representation" starkly highlights the questions of foundation and meaning.

The progress of scientific knowledge forces man to accept his contingency and his finiteness. This is where we touch upon moral issues. If the search for truth in scientific, philosophical, theological, and artistic disciplines is a moral choice that could be described as innate, then running the risk of looking for this truth with a radically new logic and set of concepts could be seen as a further moral choice. We can, therefore, point to new values in the scientific method today. A critical analysis of the foundations of these values leads us to the area of moral philosophy.

An Initial Decision in the Scientific Method: Constructing Meaning on the Basis of Nonmeaning

The diagram of the triangle to illustrate complementarity (with endless possibilities for new levels of comprehension of reality, since antagonism is never resolved at point T) illustrates the withdrawal of foundation as already mentioned in light of the work by Ladrière. There is the "undecidability." Reason can rely on nothing but itself and, at the same time, experiences its own finiteness. Reason cannot be complete; something is beyond our knowledge.

Hence comes the initial decision of a subject: to construct meaning from nonmeaning. We have a good example of this with complementarity that aims to combine antagonisms depending on their levels of reality. This decision is an essential point in scientific reasoning as illustrated by Einstein's sentence quoted above. Einstein speaks of a belief that takes us into the realm of ethics. The decision to construct meaning from nonmeaning can lead to the level of ethics according to the corresponding intentionality (personal decision) according to the commitment linked to this decision.

The Search for Meaning from Nonmeaning

It is in the search for truth that people from different disciplines (scientists, philosophers, artists, theologians) find themselves engaged in a moral choice that consists in finding the possibilities of meaning against what often appears to be a background of nonmeaning (the example of the importance of antagonisms). Every time thinking comes up against reality and bares its finiteness to represent it, there appears a basic dynamic for this reason that renders it capable of accepting new structures and building new concepts likely to favor progress in the intelligibility of reality. In this dynamic of reason, the choice of intelligibility of the world is central.

Moreover, as we have already seen, the conceptual means chosen to make progress in this intelligibility also constitute risky choice (for example, positively accepting incompleteness at the same time that the allure of completeness is still dominant). This reasoning has some link to the concepts of good and bad. To advocate certainty (or its opposite, uncertainty) is seen as positive or negative according to the individual. It becomes a question of moral commitment, of ethical decisions. Besides, the clashes of different schools of thought in each discipline serve to highlight opposing points of view that in science, for example, are of an ethical as well as a technical nature (see, for example, the debates on Darwinism and the theories of evolution).

In Ladrière's discussion of the dynamics of reason, he shows that it is founded on a prior ethical consideration.[18] The essential point is defined by the movement toward a moral life, starting from a continuous search for new representations of reality and the acceptance of their existence. Reason is seen as a representational activity that exists to analyze and understand the world. The necessary point of departure for this reasoning is the acceptance of a fundamental otherness, constituted mainly by that which resists our representations. There are moments in scientific research when reality manifests itself in a way that shows up the inadequacy of our modes of representation. We must, therefore, accommodate this "new representation."

This "acceptance" contributes in turn to realizing the knowing subject and the good scientist. The effect of this acceptance on the subject is an important element of the moral process. It is through perceiving that which "I am not" that I become myself as a subject. This otherness is not in itself a moral value, but it corresponds to a decision-making process that involves both recognition of otherness and an inclination toward unity. It is the openness to that which is other (thing and person) that falls within the realm of ethics. A new relationship with totality is initiated; this engenders a creative process that presupposes an openness to universality.

According to Ladrière, not only does everyone receive the totality of the universe through his or her personal creativity, but this creativity itself produces a new space for communication that surpasses prior inconsistencies.

All objectivity is, therefore, the external projection of that which takes place on a practical plane, whereas each practical plane is crossed, in its own right, by the demands of its own externalization. When one wants to understand the dynamics of the link between an objective and a practical plane, reason can, in a third instance, discover on the one hand, that in all these constituent objectivities bound only by their external constraints, the effect of its own activity as part of the process, and on the other that this activity can only find self-discovery through the objective status that it has given itself.[19]

Instead of considering practical human activity as a straightforward consequence of a subconscious process to be seen in the context of time and space (moreover, this activity is already an integral part of the process), the opposite is also true, according to Ladrière. The operations of this subconscious process become evident in this human activity. We therefore consider morality to be a process, whereby the subject interacts with and becomes the creator of the otherness of a totality perceived as external.

Bachelard inaugurated a movement to reconcile the spirit of contradiction and scientific thought; complementary thinking expanded the movement. Pascal's statement summarizes this well: "It is necessary to have two opposing arguments. Without that there is no understanding and everything is heretical. For every truth we always remember the opposite truth."[20] Bohr's treatment of the principle of complementarity shows us that the complementarity of antagonisms is a product of the activity of mind whereby the complexity of reality is rendered progressively more intelligible, with identity and otherness playing a tug of war. This perspective of the spirit in action takes on a moral dimension because it decides to create sense out of nonsense, meaning out of nonmeaning, and derives meaning from "nonsensical" facts while being aware of otherness and universality.

All these points of view, based on the recognition of the unity of antagonisms (or which lead to such recognition), stem from "first-time experience," that of the link between subject and the reality to which the subject belongs, the link between the uniqueness of the subject and the multiplicity of the reality in which the subject acts. All this serves to illustrate the creative process that Ladrière talks about, Weil's position on the "search for the universal," or those of Levinas on the "role of initial tension as a way of being receptive to the other."[21] We can now discuss these points of view, which should help us to discover more about the foundations of the complementarity theory.

Weil's State of the Search toward the Universal

For Weil, a Kantian, strongly influenced by Hegel, there are many other perspectives from which to consider complementarity. In *Logic of Philosophy*, notably in the chapters "Non-Meaning," "Conditions," "Absolute," and "Work," he shows how philosophy is about a personal search for meaning in life and how it identifies the problems along the way that make this search difficult if not impossible. Weil identifies in man the finiteness of the knowing subject, incapable of comprehending reality without artificially constructing it, and his infinite liberty leading him to create a meaning through the rejection of violence seen as the refusal of a coherent discourse. Philosophy is about the making of a coherent discourse, one that makes sense, which is based on knowledge (historical, political, economic, etc.) that has influenced man's attitude in the past and present.

Philosophical discourse as a rejection of violence relies on a premise (the condition, our situation in the world) that may itself appear to make no sense. Weil distinguishes between discourse and language, noting that the latter falls under the heading of "the condition," in the sense of an irreducible finiteness. It is important to insist on Weil's fundamental distinction between language and discourse. When man uses language, he uses the language of a community, not the language of the "man in a specific condition." The discourse is a search for coherence that will allow the rediscovery of a universality lost in the condition. It is worth noting here that there is a problem with scientific reasoning that sets out to describe a reality that it has only partial access to. In the process, the use of a classical language (in the case of quantum physics, for example) invariably leads to contradictions.

According to Weil, philosophical discourse is based on the premise of existence that does not appear to have any foundation and is, therefore, without meaning.

Inquiry shows how the manifestation of consciousness exists between meaning and non-meaning, both of which are constantly part of the discourse. For our present purposes it suffices to remember such opposites as language-condition, decision-situation, me-world. We can say that truth is a domain (condition, situation, world) and everything occupies this domain, revealing to us its existence, its non-meaning.[22]

Philosophical discourse as a rejection of violence is, therefore, based on a domain (condition, situation, world) that itself becomes meaningless through the act of grasping the domain. But even before this non-meaning of the domain can be thought of as such in philosophical discourse, it is first perceived as an incontrovertible fact of "the gift of life."

"Lost" universality can only be rediscovered or touched through interiority and effective action. It is by such an action in the historical world that man can understand himself and, in so doing, enter a philosophical logic by looking for total coherence with the values he has recognized through thought (we find here something of the creative process described by Ladrière). It is through this process that elevation to the universal occurs, since

> once the choice in favor of a coherent discourse has been made, the universal precedes the individual, not only in the transcendental sense but also in the most banal historical sense. Man is an individual first and foremost for the others; he does not begin by being an individual for himself.[23]

It is this elevation to the universal that confers value on all personal acts and which is the criteria for true moral philosophy for humanity. As Weil stresses, "reason is not circular." It is something that is experienced in the absence of meaning. It is a sign of the finiteness of human knowledge, of an "incompleteness," as scientists would tend to say today. It is the action that accepts finiteness, the contingency of man, which opens the way to the universal. Underlying this process is the moral choice of coherent discourse (in this case, as a way to reject violence).

This moral choice is not dissimilar to Einstein's—and many others'—belief that the world is intelligible! At the same time, something still eludes us. The subject must derive meaning from nonmeaning by accepting the limits of reason and rediscovering universality through an action, a positive choice; this is the basis of the complementarity and of the structure of different levels of reality we have discussed in reference to Bohr and Nicolescu. It is this elevation to the universal that confers validity on all personal action and which, according to Weil, is the only criteria for a true moral philosophy for humanity.

Weil's incisive analysis allows us to recover the distinction between the different levels on which we work. Here, it is the rejection of violence that allows us to pass from the metaphysical level (search for meaning from nonmeaning) to the level of moral philosophy (the subject finds meaning by rediscovering universality through action, a choice, in effect, which implicates him). As we have seen, this action contains an acceptance of finiteness and the contingence of man. Such "wisdom" (learn from man's contingence) provides a privileged space for dialogue with theologians.[24]

We have referred to the works of Weil and Ladrière in our discussion of the foundations of complementarity. This appears to be an illustration among other things of the problem of Sameness and Otherness. What is particularly interesting is that our approach, which started out by thinking about the current evolution of ideas in science, in fact leads us to moral philosophy by way of metaphysics (three quite different areas).

The Meaning of Mystery

The arguments we have presented so far can also be described in terms of a dialectic of mystery. What mystery are we referring to? It is the "mystery of knowing" that has been our theme until now, emanating from a discussion on the evolution of scientific knowledge. Einstein's assertion that "the most incomprehensible thing about the world is that it is comprehensible" and the demonstration of "fecundity" of the idea of incompleteness are like two "signs" to the mystery of knowing in modern scientific reasoning.[25]

One of the most interesting ways of rethinking the concept of mystery in the twentieth century was proposed by Gabriel Marcel.[26] He criticizes philosophers for "abandoning" mystery to theologians and popularizers. Marcel not only considers the mystery of knowing but also the mystery of the union of body and soul, and the mysteries of love, hope, presence, and being. In respect to the questions we are concerned with here, the most interesting aspect is the distinction he makes between problem and mystery. The problem is a question that we ask ourselves about elements that have been laid out before us, as it were, and that are, generally speaking, external to us. Of course, if we think about it, we have to acknowledge that there is always the link of knowing between them and us. But characteristic of this form of thinking that considers problems is the implicit postulate that the fact of knowing does not redefine the problem. Moreover, apart from the purely intellectual interest we might have in them, there are no negative repercussions on us. The problems of classical mathematics are the most obvious example of this. There is mystery, on the contrary, when the one asking the question belongs to the very thing about which he is asking the question, that is, the mystery of being, about which I can only inquire into insofar as I am.

A mystery is a problem which encroaches on its own data, . . . it is a problem that steps on its own immanent conditions of possibility. Or else: mystery is something I find myself caught up in and, I would add, not in a partial way by some predetermined or specialized aspect of myself, but on the contrary completely, since I constitute a unit which by definition can never quantify itself and which can only be an object of creation and faith.[27]

Mystery, therefore, breaks down the barrier between the "in me" and the "before me" that characterizes the domain of problem solving, even if we know that the act of knowing is an intercession and that one can never attain an "in one's self." There is mystery of being that is also "the mystery of the act or of thought, which can also be translated as follows: we cannot ask ourselves about being as if the thinking that asks about being was outside of being." There is mystery of knowledge: "Knowledge depends on a mode of participation which no epistemology can hope to account for since it is itself the source of enquiry."[28]

For Marcel, mystery is neither the unknowable nor a sort of pseudo-solution. Far from being a "knowledge gap," mystery is a call for exploration. This rehabilitation of mystery at a philosophical level (G. Marcel employs the term "meta-problematical" to describe mystery) allows for an interesting bridge with theology, as I have analyzed in my book. This is close to the approach of Saint Augustine, who said in another context that mystery is not what one cannot understand but what one will never cease to understand.

With Marcel's view in mind, let us now return to the questions of incompleteness, complementarity, and the logic of antagonisms. This is an example of the "mystery of knowing." The model relating to the concept of levels of reality expresses the mystery of knowing that the scientist is faced with. In science, we can also talk about the involvement of a thinking subject (man is a part of the nature he analyzes) even if the scientist's commitment is not as strong as the philosopher's, as defined by Marcel. We can even talk about the alteration of reality by the subject who is analyzing it even if, once again, the alteration is not as strong as in the philosophical question of being, as described by Marcel (the subject in physics is not personalized, the alteration of reality introduces itself by the measuring operations that itself is depersonalized).

Nonetheless the question of knowing in modern science refers the scientist to the mystery of knowing as so well expressed by Einstein. Hence, the search for the unity of antagonisms harks back to a "first experience" which is that of the link between the subject and the reality to which it belongs (the link between the unicity of the subject and the multiplicity of the reality in which it operates). The acceptance of the mystery of knowing is once again linked to the finiteness of man: it involves an implicit and explicit moral choice depending on the scientists! In the case of complementarity, incompleteness and the concept of the level of reality, we can talk more in terms of the "dialectics of mystery" in sciences.[29]

Conclusions: Opening Ways to the Mystery of Man

Twentieth-century science leads the scientist to ask about man's place in the history of the universe. This question arises out of thinking about the foundations of the major theories and the underpinnings of scientific reasoning. Classical science provided us with very mechanistic diagrams for representing the world, defining it as a large clock in which man is seen as a simple cog in this contraption. We have been able to measure the influence of scientism in order to better discern what consequences the changes in perspective of contemporary science can have today on society's mentalities and ethics.[30]

"Scientific" objectivity, brandished as a supreme criteria for truth, has had a far more profound effect than scientists could ever have expected. *The subject has become the object.*[31] That man should be the "object" of knowledge is perfectly normal for the scientist. It is unacceptable, however, that in the name of scientism he becomes the object of exploitation, ideological experiences, or scientific experiments, to be dissected, standardized, manipulated. Of course, this was not the objective of the majority of scientists who tried to establish scientific objectivity. What comes out of this is the "moral influence" that ideas and concepts that originate in science can have on society. In the sciences of the universe and matter, the subject has been partially reintegrated via the acceptance that it is linked to the object. The vision of an "uncertain world," in the words of scientists such as d'Espagnat and Prigogine, calls us to go beyond scientific materialism, even if there is strong reticence among biologists.

The withdrawal of foundation discussed earlier reinforces this view. We must be careful, however, not to fill this uncertainty with a more-or-less disguised return to old certainties. The temptation to fill in the gaps of Gödel's incompleteness by a "God of the gaps" is just one example. Let us allow man to receive reality as it presents itself to us; let us give rein to reason that will be open to all eventualities, to be able to articulate the unicity of man and the multiplicity of reality. The mystery is not of the order of magic; it is of the order of intelligence that progresses without ever being self-sufficient.

Let us enter such a world. To find the meaning of this otherness and of this fundamental unity between the subject and reality is to make the choice in an uncertain world of positing, the possibility of an intelligibility, the existence of a meaning. To accept otherness, to avoid simplifying the complex, to think differently, this is what the scientist must choose—a moral choice, reflecting on the mystery of man in nature. In this way, fundamental moral attitudes can be called upon in all search for the truth, notably in science. We must have honesty in this search for truth, of course, acknowledgment of the foundation of meaning where human reason cannot come full circle, active acceptance of the incompleteness of all knowledge, and a dialectic approach whereby something will always elude us. We have to enter into an acceptance of a fundamental otherness for the subject (otherness looking for a link with unity), acceptance of a finiteness and of the contingence of the knowing subject, and the choice of finding meaning from nonmeaning. A certain humility will result, proof of progress of knowledge that will see the abandonment of definitive certainties for an incompleteness that does not deny the search for truth but displays our own incapacity to reach it on our own, while making us more open to the importance of this truth. All this is covered by moral philosophy! It is on this note that it is interesting to assess the relationships between the scientist and the believer in their quests.

NOTES

1. Philippe Franck, *Einstein* (Paris: Albin Michel, 1968), 126

2. François Jacob, *The Game of Possibilities* (Paris: Fayard, 1981).

3. Ibid., 28.

4. Ibid., 119.

5. See, in particular, Ilya Prigogine and Isabelle Stengers, *La Nouvelle Alliance* (Paris: Gallimard, 1979).

6. See Bernard d'Espagnat's chapter in this volume and also Bernard d'Espagnat, *Une incertaine réalité* (Paris: Gauthier-Villars, 1985).

7. Wittgenstein quoted in Michel Simon, *Penser et croire au temps des sciences cognitives* (Grenoble: Archives contemporaines, 2001).

8. Douglas Hofstadter, *Gödel, Escher, Bach* (Paris: Interéditions, 1985), 153.

9. Jean Ladrière, *L'Abîme, in Savoir, faire, espérer; les limites de la raison,* ed. J. Beaufret (Bruxelles: Pub. Facultés Univ. St Louis, Tome 1, 1976), 171–91.

10. Ibid., 175.

11. Ibid., 177.

12. Gerard Holton, *L'imagination scientifique* (Paris: Gallimard, 1991).

13. W. Heisenberg, *La Partie et le Tout* (Paris: Albin Michel, 1972), 144.

14. Letter from Bohr, quoted by M. Jammer, *The Philosophy of Quantum Mechanics* (New York: J. Wiley and Sons, 1974), 54.

15. Stéphane Lupasco, *L'expérience microphysique et la pensée humaine* (Paris: PUF, 1941); idem, *L'homme et ses trois éthiques* (Paris: Le Rocher, 1986); idem, *Nous, la particule et le monde* (Paris: Le Mail, 1985); and idem, *Science, Meaning and Evolution* (New York: Parabola Books, 1991).

16. Basarab Nicolescu, *La Transdisciplinarité* (Paris: Le Rocher, 1996), 25.

17. Edgar Morin, *Science et conscience* (Paris: Fayard, 1982).

18. Jean Ladrière, *L'éthique et la Dynamique de la Raison,* in Rue Descartes n°7, *Logiques de l'éthique* (Paris: Albin Michel, 1993).

19. Ibid., 58.

20. Blaise Pascal, *Pensées.*

21. Eric Weil, *Logique de la philosophie* (Paris: Vrin, 1950); Emmanuel Levinas, *Totalité et Infini,* (Paris: Kluwer Academic, 1971).

22. Ibid., 95.

23. Ibid., 68.

24. See my *Entre Science et Religion* (Paris: Le Rocher, 1998).

25. Franck, *Einstein*, 321.

26. Gabriel Marcel, *Positions et approches concrètes du Mystère ontologique* (Paris: Nauwelaerts et Vrin, 1949); idem, *Etre et Avoir* (Paris: Aubier, 1935), 183; idem, *Les Hommes contre l'humain*, (Paris: La Colombe, 1951), 69.

27. Marcel, *Etre et Avoir*, 183.

28. Marcel, *Positions et approches concrètes du Mystère ontologique*, 101.

29. Marcel, *Les Hommes contre l'humain*, 69.

30. Basarab Nicolescu, *La Transdisciplinarité* (Paris: Le Rocher, 1996), 24.

31. See the chapter "The Object" in Weil, *Logique de la philosophie*.

PART IV

Agreements and Conflicts between the Two

❧

Modern Cosmology and the Quest for Meaning
A Dialogue on the Road to Knowledge

BRUNO GUIDERDONI

According to a commonly acknowledged idea, science deals with *facts*, whereas religion deals with *meanings*. If science attempts to answer the "how" and religion the "why," there should not be any conflict between the two. Unfortunately, the situation is not so simple. It is true that science deals with *efficient* causes and religion with *final* causes, to use the technical words of Aristotelian philosophy.[1] But the general trend in the development of sciences is that the efficient causes push the final causes backward and eventually eliminate them.

This progressive replacement of the explanation in terms of final causes by the explanation in terms of efficient causes has been happening in the West since the Renaissance. In the Middle Ages, Jews, Christians, and Muslims shared the same vision of the world, even if there were already long-lasting controversies and hot debates on cosmological issues. The men and women of faith of the Middle Ages did not see only things and phenomena around them; they primarily contemplated symbols and looked for spiritual unveiling through their study of the cosmos. The epoch of the medieval synthesis between the Aristotelian-Ptolemaic cosmology and the teachings of the Holy Scriptures has passed away, and the development of modern science has led to a profound spiritual crisis in the West. Man has lost his central place in the cosmos and has been rejected onto a standard planet orbiting a standard star in a standard galaxy located somewhere in the dull immensity of space. Such a science is value-neutral and devoid of any meaning. In the words of Claude Levi-Strauss, "the world began without the human being, and will end without him."[2]

The conflict between science and religion ceased in the West when religion admitted that it has nothing to say on cosmology. The fields simply do not overlap because science has colonized the whole of "reality." To do so, it has defined reality as being only what can be studied scientifically. Theologians now have to explain why God appears to be hidden under the thick curtain of phenomena. Ideas such as those of *kenosis* and *tsimtsum*, which flourished respectively in Christian and Jewish theological thinking, have undergone a fascinating revival, and are now used by these theologians to explain why God retires, apparently, to let the cosmos be ruled by its own laws, without any sign of direct divine intervention. The emphasis is put on the (relative) independence granted by God to the laws of nature and the (relative) freedom granted by God to man.

As is well known, the Islamic tradition has taught that God is nearby and continuously acts in creation. "Each day some task engages him" (Qur'an 55:29). So Muslim theologians are unable to follow the path of some of the Western theologians in the direction of a *Creator* who would let his creation behave by itself with so much independence that he finally becomes a new kind of *Deus otiosus,* either by will or by experience of human weakness. God is hidden, but he is also apparent, according to his beautiful names *azh-Zhahir wa-l-Batin.* The Creator is so great that his creation has no flaw. But he is also apparent in and through his creation.

The fundamental mystery that subtends physics and cosmology is the fact that the world is intelligible. For a believer, the world is intelligible because it is created. The Qur'an strongly recommends pondering and meditating upon creation to find the traces of the Creator in its harmony. Hence, the so-called cosmological verses are frequently quoted as one of the intellectual miracles included in the Qur'an:

In the creation of the heavens and the earth, and in the alternation of night and day, there are signs for men of sense; those who remember God when standing, sitting, and lying down, and reflect on the creation of the heavens and the earth, saying: "Lord, You have not created this in vain. Glory be to you! Save us from the torment of the fire." (3:190–91)

The exploration of the world is encouraged, provided the explorer is wise enough to acknowledge that the harmony that is present in the cosmos originates in God.

By looking at the cosmos, the intelligence God has put in us constantly meets the intelligence he has used in creating the things. The Qur'an mentions the regularities that are present in the world: as well as "you will find no change in God's custom" (Qur'an 35:43), "there is no change in God's creation" (Qur'an 30:30). Clearly, this does not mean that the creation is immutable but that there is a "stability" in the creation that reflects God's immutability. The reader's attention is also drawn to the "numerical aspect" of cosmic regularities. The Qur'an says: "The Sun and the Moon [are ordered] according to an exact computation [*husban*]" (55:5; see also 6:96, 10:5, 14:33). So a Muslim cosmologist is not surprised that the laws of physics we design and use to describe cosmic regularities are based on mathematics.

We live in a very peculiar epoch for the understanding of the structure and history of the cosmos. In the last decades, there have been spectacular breakthroughs mainly due to the extraordinary development of observation techniques. As a consequence, we have acquired a treasury of images we are the first generation to contemplate—the image of the Earth in the darkness of the sky, the wide diversity of appearance of the surface of other planets and satellites in the solar system, the mapping of our galaxy at all wavelengths, the discovery of very energetic phenomena such as star explosions, or the potential census of billions of distant galaxies in deep surveys. We now have access to distances, epochs, and structure sizes that were simply unthinkable at the epoch of the Middle Ages when the Arab astronomer al-Farghani computed the distance to God's throne from the assumptions of the Ptolemaic cosmology, and found a value of 120 million kilometers.[3] These new images have deeply changed our awareness of the cosmos.

To understand the structure of the universe, cosmologists must track its history, one theoretically reconstructed from the data by means of elaborated mathematics. No doubt there are many bold speculations and crazy ideas in the interpretation. But reality resists, and not all theories are in agreement with the facts. On the contrary, the standard theory now appears as a powerful tool to guide new discoveries. To cut a long story short, cosmologists now think that the universe is expanding, and that the expansion phase started from a dense, hot stage called the big bang. During the expansion, the matter/radiation content of the universe dilutes and cools, and the relative abundance of various species of elementary particles changes. About 100 seconds after the big bang, light nuclei begin to form. About 300,000 years after, the universe becomes neutral and transparent, and the light emitted by the so-called last-scattering surface at that epoch is observed as the 2.728 K blackbody radiation of the Cosmic Microwave Background. The story is now well documented, but there are several topics for which our incapacity to solve recurrent puzzles probably points to the metaphysical structure of reality. In the following, I would like to briefly address two of these puzzles.

The first puzzle deals with fine-tuning in structure formation. Regions that are separated by more than about one degree on the last-scattering surface have never been in causal connection before and should have widely different temperatures, in contrast with the remarkable isotropy that is actually measured: this is the so-called isotropy problem. Moreover, the density of the universe is close to unity, and the spatial geometry is almost flat, whereas all values for the density parameter are a priori possible: this is the so-called flatness problem. As a result, our observable universe seems to have emerged from a very peculiar set of initial conditions. In parallel, it is now clear that these patterns are necessary conditions for the appearance of complexity in the universe.

For instance, a very large density parameter would have produced a fast collapse in a time scale much lower than the stellar lifetimes that are necessary for the chemical enrichment of the interstellar medium (and the subsequent formation of planets), whereas a very low density parameter would have resulted in a very diluted universe, with low mass structures that are unable to retain their gas. Of course, a philosophical explanation in terms of final causes can be introduced to give meaning to this type of fine-tuning (and other cosmic coincidences gathered under the term of *anthropic principle*).[4] It can be divine intervention in a religious prospect, or a natural trend of matter toward self-organization in a pantheistic prospect. But this is unacceptable for modern science. As a matter of fact, the elimination of explanations in terms of final causes is at the heart of the development of cosmology. The current explanation of the isotropy and flatness problems (and other related puzzles) is that the universe has undergone a stage of exponential inflation that has inflated a small, causally connected patch beyond the size of the observable universe, and has erased spatial curvature. This explanation avoids the introduction of any argument on final causes about the set of initial conditions the universe started from.

By the same token, the origin of the inhomogeneities that will produce the large-scale structures after gravitational amplification is explained by inflation; they are simply quantum fluctuations inflated to macroscopic scales. The problem is that the current theory is not able to predict the amplitude of these fluctuations, which is measured at the relative level of one part in 100,000 ($Q=10^{-5}$) on the last-scattering surface. When a complete theory of inflation emerges, it will have to predict this value, which now appears only as a free parameter. But it is already clear that this value is also a necessary condition for the appearance of complexity in the universe. With $Q=10^{-6}$, gas cannot cool in the potential wells of haloes and no stars can form. With $Q=10^{-4}$, galaxies are so dense that frequent stellar encounters hamper the existence of stable planetary orbits, which are a necessary condition for the existence of living ecosystems that draw their energy from stellar radiation. Again, our observable universe seems to have emerged from a very peculiar set of initial conditions.

Cosmologists have a new theory that avoids the introduction of final causes called *chaotic inflation*. In chaotic inflation, inflation eternally takes place and makes new patches of exponentially inflating space-time that causally decouple one from the other. Subsequently, the inflationary stages turn into the normal expansion phases. In this context, the laws and constants of physics are fixed by symmetry breaking and getting different values in the different patches. Consequently, with an infinite number of realizations, we must not be surprised that there is at least one patch of the universe that has the values of the laws, constants, and of Q suited to the appearance of complexity. The question of knowing whether this theory is testable is still open. But it is not our concern here.

At the current stage of explanation, the apparent fine-tuning in the universe is not due to a peculiar set of initial conditions but to the exploration of a range of possible values in various patches of the universe. We simply live in a patch that has values suited to the existence of complexity. But this type of explanation ignores the "power" allotted to the principles of quantum mechanics and the fundamental field theory. When an over-arching field theory is developed (maybe some kind of superstring theory), it will turn out that it has the possibility of generating patches where complexity is possible. So we shall have to push our explanation forward again to a broader theory, a quest that appears endless. The irony is that, when cosmologists try to evacuate final causes, they make new theories and discover new phenomena but always face the same type of puzzle. The existence of fine-tuning in the universe surely tells us something about reality. But what? Man can readily understand that it is a divine sign; if he does not, the door is open to an endless exploration of the cosmos that displaces and magnifies the puzzle, until he finally acknowledges it. "Whichever way you turn, there is the Face of God" (Qur'an 2:115).

The second puzzle deals with the universality of the laws. Some cosmologists use the word "universe" for each of these causally disconnected patches and the word "multiverse" to name the ensemble of all these patches generated by chaotic inflation. Of course, there is some ideology in the choice of the names. According to its symbolical etymology, the universe is a sign that is directed "toward the One" (*unum versus*). Do many worlds suggest many gods? In any case, in the mind of some of those who promote the multiverse, new cosmology seems to be more sympathetic with polytheism than with monotheism. However, all these patches of the universe are actually linked by the fact that they are ruled by the same principles of quantum physics and the same overarching field theory. For that reason, there is actually a *single* universe. Why are the laws of quantum physics so *universal*?

Here again, modern cosmologists do not wonder enough about the continuous validity of the laws. There has been a long controversy in Islam on the existence and status of the secondary causes. It is well known that the Ash'arite theology strongly questions the very existence of causality. The position is that there are no secondary causes simply because God, as the "primary" cause, does not cease to create again the world at each instant. In this continuous renewal of creation (*tajdid al-khalq*), the atoms and their accidents are created anew each time. As a consequence, the regularities observed in the world are not due to causal connection but to a constant conjunction between the phenomena, which is a custom established by God. This principle of Islamic theology should be primarily understood as an emphasis on a metaphysical mystery, the continuous validity of the laws. God's permanence makes creation behave regularly in spite of the continuous renewal. "You will not see a flaw in the Merciful's creation. Turn up your eyes: can you detect a single crack?" (Qur'an 67:3).

The renewal of creation taught by Islamic doctrine also means the continuous appearance of new creatures. According to the views of the Akbarian school, founded upon the work of Muhyi-d-din Ibn Arabi, who died in 1240, the creation is God's self-disclosure to himself through the veils and signs of the creatures. The things "are" not, since only God is. They only own a given preparation to receive being and qualities from God. As a consequence, since the status of the cosmos is paradoxical, between absolute Being and absolute nothingness, we cannot expect to reach clear-cut statements about the fundamental reality of the world.

The ultimate reality is hidden and our descriptions will always be approximate.

God is infinite and "self-disclosure never repeats itself."[5] So God's
self-disclosure is endless. At each level of the cosmos, there are always
new things continuously "poured" into disclosure. What appears in the
creation exactly corresponds to the flow of possible things. This is why,
according to the great theologian and mystic al-Ghazali, who lived in the
eleventh century, "There is nothing in possibility more wondrous than
what is,"[6] because what is actually reflects God's desire to show God to
us. This helps us understand the prophetic saying: "Curse not time, for
God is time."[7] After all, the production of an infinite number of "patches" of the physical universe described by chaotic inflation could fit in
this view of God's eternal self-disclosure. The appearance of "emerging
properties" at all levels of complexity, and particularly the appearance of
life and intelligence, is another aspect of this continuous self-disclosure.
This is why Ibn Arabi comments, "God does not become bored that you
should become bored."[8] We cosmologists surely understand this, since
we are continuously astonished by the beauty of the phenomena unravelled by our new observing tools.

The appearance of the human being was made possible by many
anthropic "coincidences" in the laws of physics and the values of constants, which fix the properties of the cosmic and terrestrial structures.
The extension of time behind us and of space around us is a necessary
condition for our existence, as the vast extensions of the deserts of sand
and ice are necessary for the ecological balance of the Earth. But this is
of little interest in our spiritual call for an endless quest—the quest for
knowledge that is the core of our nature and dignity. However, there is a
significant difference between scientific pursuit and the spiritual quest,
which deals with the ending point of our existence. Contrary to scientific pursuit, the spiritual quest is not limited to the intellectual search
for truth and the production of useful outcomes. It primarily aims at
transforming the human so that he can be prepared for the afterlife.

Let us look at Averroes and Ibn Arabi, who happened to meet in
Cordoba, probably around 1180. Averroes, who then was already a re-

nowned philosopher, defended that human reason was able to reach all the truth accessible to the human, and not less than what was brought by revelation under the veils of the dogmas and symbols for the benefit of those who are not experts in science. Averroes heard that the young Ibn Arabi had been granted spiritual enlightenment and was eager to meet him. On their meeting, Ibn Arabi wrote:

When I entered in upon [Averroes], he stood up in his place out of love and respect. He embraced me and said, "Yes." I said, "Yes." His joy increased because I had understood him. Then I realized why he had rejoiced at that, so I said, "No." His joy disappeared and his color changed, and he doubted what he possessed in himself.

Then comes the explanation of these strange exchanges. Averroes asks the crucial question that we are interested in: "How did you find the situation in unveiling and divine effusion? Is it what rational consideration gives to us?" Ibn Arabi replies, "Yes and no. Between the yes and the no, spirits fly from their matter and heads from their bodies," and he reports on Averroes' reaction. "His color turned pale and he began to tremble. He sat reciting, 'There is no power and no strength but in God,' since he has understood my allusion."[9]

Ibn Arabi alluded to eschatology by recalling that, even if reason can go very far in its attempt to grasp reality, nobody has been intimately changed by scientific knowledge. According to the teachings of Islam, we shall have to leave this world at the moment of our death in order to pursue our quest for knowledge, and enter another level of being that is a broader locus for God's self-disclosure. The Islamic tradition promises that the quest for knowledge will end when the elects contemplate God's face on the so-called Dune of Musk (*al-kathib*) that is located on the top of the heavenly gardens, at the last frontier of creation. Religion is providentially revealed to prepare us to face Absolute Reality, which is another name of God. But the end of this quest will not be the end of knowledge. On the contrary, the elects' contemplation of God will continuously be renewed, because they will know, according to a divine saying, "what no eye has seen, what no ear has heard, and what has never passed into the heart of any mortal."[10] Our reason could estimate that this is impossible, since we do not conceive "how" this can physically happen. But indeed, the Dune is the locus of the answers to the "why" questions, without "how."

Because of our spectacular progress in the scientific understanding of the universe, we have forgotten contemplation, which is necessary to the human being. It is this kind of awareness that can help reconcile science and religion, though not by a low-level concordism. To conclude, I would like to mention three qualities that seem to be relevant for all those who, as scientists and believers, keep a continuous tension toward truth: gratefulness (*shukr*), fear (*taqwa*), and perplexity (*hayrah*). In all religious and mystical knowledge, there is gratefulness for the marvels of the cosmos, fear for the sense of transcendence it inspires, and perplexity for the continuous existence of unsolved puzzles that point at more fundamental mysteries. In the Islamic vision, gratefulness refers to the Names of Beauty (*asma' al-Jamal*), fear to the Names of Majesty (*asma' al-Jalal*) that show up in the worlds, and perplexity to the coexistence of opposite qualities that can be solved only in Allah, who is the Name of the Synthesis (*ism al-jami'*). The spiritual pursuit is not limited to the intellectual contemplation of truth, but it aims at salvation, which is the ultimate meaning of the human. Gratefulness, fear, and perplexity are three modes of the fundamental bewilderment that is produced by the contemplation of the cosmos, a bewilderment that is a way of worshiping God. Such an attitude should lead scientists to an increasing sense of responsibility in the technological applications of modern science.

NOTES

1. To the questions "Why does the Sun shine?", an answer in terms of final causes could be "It shines to give light to the Human being," whereas an answer in terms of efficient causes could be "It shines because its surface is hot."

2. Claude Levi-Strauss, *Tristes Tropiques* (Paris: Librarie Plon, 1955).

3. Edward Grant, *Planets, Stars & Orbs, The Medieval Cosmos, 1200–1687* (Cambridge: Cambridge University Press, 1994), 433.

4. J. D. Barrow and F. J. Tipler, *The Anthropic Cosmological Principle* (New York: Oxford University Press, 1986).

5. Sufi axiom quoted by William Chittick in *The Sufi Path of Knowledge* (Albany: SUNY Press, 1989), 409.

6. Al-Ghazali quoted in ibid., 409.

7. Prophetic saying (*hadith*) quoted in ibid., 107.

8. Ibn Arabi quoted in ibid., 101.

9. I use the translation in William Chittick, *The Sufi Path of Knowledge* (Albany: SUNY Press, 1989).

10. Divine saying (*hadith qudsi*).

Science and Buddhism

TRINH XUAN THUAN

Is There a Basis for Dialogue?

As an astrophysicist studying the formation and evolution of galaxies, I often face questions about matter, space, and time. As Vietnamese-born and raised in the Buddhist tradition, whenever I have come up against these concepts, I couldn't help wonder about how Buddhism would have dealt with them and how its view of reality compares to the scientific viewpoint. But I wasn't sure whether such questions even made sense. I was familiar with and appreciated Buddhism as a practical philosophy that provides a guide for self-knowledge, spiritual progress, and becoming a better human being. Thus, as far I knew, Buddhism was primarily a pathway leading to enlightenment, a contemplative approach with an essentially inward gaze, in contrast to science's outward look.

What's more, science and Buddhism have radically different methods of investigation of reality. In science, intellect and reason play the leading roles. By dividing, categorizing, analyzing, comparing, and measuring, scientists express natural laws in the highly abstract language of mathematics. Intuition is not absent in science but it is only useful if backed up by a coherent mathematical formulation and validated by observation and analysis. On the other hand, it is intuition—or rather, inner experience—that plays the leading role in the contemplative approach. Rather than breaking up reality, Buddhism strives to understand it in its entirety. Buddhism has no use for measuring apparatus and does not rely on the sort of sophisticated observations that form the basis of experimental science. Its statements are more qualitative than quantitative. So I was far from sure that there would be any point confronting science and Buddhism.

I was afraid that Buddhism would have very little to say about the nature of phenomena, because this is not its main interest, whereas such preoccupations lie at the heart of science. In the summer of 1997, I met the French Buddhist monk Matthieu Ricard at the University of Andorra where we were both giving lectures. He was the ideal person with whom to discuss these issues. He was trained as a scientist (he has a doctorate in biology from the Pasteur Institute in Paris), so is familiar with the scientific method. But he is also well versed in Buddhist texts and philosophy, as he left the scientific world about thirty years ago to become a Buddhist monk in Nepal. We had many fascinating discussions during our long walks together in the inspiring mountain scenery of the Pyrenees; our conversations were always mutually enriching. They led to new questions, original viewpoints, and unexpected syntheses that required further study and clarification, and still do so. I shall discuss below the main issues that sometimes divided us, sometimes united us. A book, *The Quantum and the Lotus*, was born from those friendly exchanges between an astrophysicist—who wanted to confront his scientific knowledge with his Buddhist philosophical origins—and a Western scientist who became Buddhist and whose personal experience has led him to compare these two approaches.[1] At the close of our conversations, I must say that my admiration for how Buddhism analyzes the world of phenomena had grown considerably. Buddhism has thought deeply and in an original way about the nature of the world.

But the ultimate goals for the pursuit of knowledge in science and in Buddhism are not the same. The purpose of science is to find out about the world of phenomena. In Buddhism, knowledge is acquired essentially for therapeutic purposes and the objective is not to find out about the physical world for its own sake but to free ourselves from the suffering that is caused by our undue attachment to the apparent reality of the external world. By understanding the true nature of the physical world, we can clear away the mists of ignorance and open the way to enlightenment. It is not my purpose in this chapter to make science sound mystical nor to justify Buddhism's underpinnings with the discoveries of science. Science is perfectly self-sufficient and accomplishes well its stated aim without the need of a philosophical support from Buddhism or from any other religion. Buddhism is a science of the enlightenment, and whether it is the Earth that goes around the Sun or the contrary cannot have any consequence on its philosophical basis. But because both are quests for the truth and both use criteria of authenticity, rigor, and logic to attain it, their respective views of the world should not result in an insuperable opposition but rather to a harmonious complementarity. Werner Heisenberg expressed this eloquently: "I consider the ambition of overcoming opposites, including also a synthesis embracing both rational understanding and the mystical experience of unity, to be the mythos, spoken or unspoken, of our present day and age."[2]

I shall discuss below the Buddhist concepts on interdependence, emptiness, and impermanence and how they find an echo in modern science. I shall outline how Buddhism rejects the idea of an "anthropic" principle. Finally, I conclude with the idea that science and spirituality are two complementary modes of knowledge and that must go hand in hand so we do not forget our humanity.

Interdependence

Buddhism and the Interdependence of Phenomena

One of Buddhism's central tenets is the interdependence of phenomena: nothing exists inherently or is its own cause. An object can be defined only in terms of other objects and can exist only in relationship to others. In other words, *this* arises because *that* exists. Interdependence is essential to the manifestation of phenomena. In Buddhism, the perception of distinct phenomena resulting from isolated causes and conditions is called relative truth or delusion. Our daily experience makes us think that things possess a real, objective independence, as though they existed all on their own and had intrinsic identities. But Buddhism maintains that this way of seeing phenomena is just a mental construct. It adopts the notion of mutual causality; an event can happen only because it is dependent on other factors. Any given thing in the world can appear only because it is connected, conditioned, and in turn conditioning. An entity that exists independently of all others, as an immutable and autonomous entity, couldn't act on anything or be acted on itself. Buddhism thus sees the world as a vast flow of events that are linked together and participate in one another. The way we perceive this flow crystallizes certain aspects of the nonseparable universe, thus creating the illusion that there are autonomous entities completely separate from us. Thus phenomena are simply events that happen in some circumstances. This view does not mean that Buddhism denies conventional truth—the sort that ordinary people perceive or the scientist detects with his apparatus—or that it contests the laws of cause and effect or the laws of physics and mathematics. It simply holds that, if we dig deep enough, there is a difference between the way we see the world and the way it really is.

The most subtle aspect of interdependence concerns the relationship between a phenomenon's "designation base" and its "designation." An object's designation base refers to its position, dimension, form, color, or any other of its apparent characteristics. Together, these characteristics form the object's designation—a mental construct that attributes an autonomous distinct reality to that object. In our daily experience, when we see an object, we aren't struck by its nominal existence but by its designation. Because we experience it, Buddhism does not say that the object doesn't exist. But neither does it say that the object possesses an intrinsic reality. So it concludes that the object exists (thus avoiding the nihilistic view that Westerners too often attribute mistakenly to Buddhism) but that this existence is purely interdependent. A phenomenon with no autonomous existence, but that is nevertheless not totally inexistent, can thus act and function according to the laws of causality.

Nonseparability in Quantum Mechanics

A notion strikingly similar to that of Buddhism's interdependence is the concept of nonseparability in quantum mechanics, which is based on the famous thought experiment proposed by Einstein, Podolsky, and Rosen (EPR) in 1935. In simplified terms, the experiment goes as follows.

Imagine a particle that disintegrates spontaneously into two photons, *A* and *B*. The law of symmetry dictates that they will travel in opposite directions; if *A* goes northward, then we will detect *B* to the south. It all seems perfectly normal. But that's forgetting the strangeness of quantum mechanics. Particles can also appear as waves. Before being captured by the detector, *A* appeared as a wave not a particle. This wave was not localized, so that there is a certain probability that *A* might be found in any direction. It's only when it has been captured that *A* changes into a particle and "learns" that it's heading northward. But, if *A* didn't "know" before being captured which direction it had taken, how could *B* have "guessed" what *A* was doing and adjust its behavior accordingly so that it could be captured at the same time in the opposite direction? This is impossible, unless *A* can inform *B* instantaneously of the direction it has taken. But as Einstein said, "God does not send telepathic signals," and there can be "no spooky action at a distance."[3] He therefore concluded that quantum mechanics did not provide a complete description of reality, that *A* must "know" which direction it was going to take and "tell" *B* before they split up. According to him, there must be hidden variables and quantum mechanics must be incomplete. And yet Einstein was wrong.

In 1964, John Bell devised a mathematical theorem called Bell's Inequality, which could be verified experimentally if particles really did have hidden variables. In 1982, Alain Aspect carried out a series of experiments on pairs of photons and found that Bell's inequality was always violated.[4] Quantum mechanics was right and Einstein was wrong. In Aspect's experiment, photons *A* and *B* were 12 meters apart, yet *B* always "knew" instantaneously what *A* was doing and reacted accordingly. In the latest experiment carried out by Nicolas Gisin, the photons are separated by more than 10 km, and yet their behaviors are perfectly correlated.[5] This is bizarre only if, like Einstein, we think that reality is cut up and localized in each photon. The problem goes away if we admit that *A* and *B*, once they have interacted with each other (the physicists describe them as entangled), become part of a nonseparable reality, no matter how far apart they are, even if they are at opposite ends of the universe. *A* doesn't need to send a signal to *B* because they share the same reality. Quantum mechanics thus eliminates all idea of locality and provides a holistic view of space. The notions of "here" and "there" become meaningless because "here" is identical to "there," which is what physicists call nonseparability. So phenomena do seem to be interdependent at the subatomic level, to use the Buddhist term.

Foucault's Pendulum and Interdependence in the Macrocosm

Another fascinating and famous physics experiment shows that interdependence isn't limited to the world of particles but applies also to the entire universe. This is the pendulum experiment carried out by Léon Foucault in 1851 to demonstrate the rotation of the Earth. We are all familiar with the behavior of the pendulum. As time passes, the direction in which it swings changes. If the pendulum were placed at either the North or South Pole, it would turn completely round in twenty-four hours. Foucault realized that, in fact, the pendulum always swung in the same direction and it was the Earth that turned.

But there remains a puzzle not clearly understood to this day. The pendulum is attached to a building that is attached to Earth. The Earth carries us at some 30 km/s around the Sun, which is itself flying through space at 230 km/s in its orbit around the center of our Milky Way Galaxy. Our galaxy, in turn, is falling toward the Andromeda Galaxy at 90 km/s. The Local Group of galaxies, whose most massive members are the Milky Way and Andromeda, is moving at 600 km/s under the gravitational attraction of the Virgo cluster, and of the Hydra-Centaurus supercluster. The latter is itself falling toward the Great Attractor, the mass of which is equivalent to that of tens of thousands of galaxies. All of these masses and motions are local. Yet, the Foucault pendulum seems to disregard all of them and align itself with the rest of the universe, that is, with the most distant clusters of galaxies known. Thus, what happens here on Earth is decided by all the vast cosmos. What occurs on our tiny planet depends on all the structures in the universe. Why does the pendulum behave in that way? We don't know.

Ernst Mach thought that it could be explained by a sort of omni-presence of matter and its influence.[6] According to Mach, the correla-tion between the plane of oscillation of the Foucault pendulum and the distant clusters of galaxies comes from the distant universe being responsible for the inertia of the pendulum and, hence, of its motion through a mysterious interaction that he did not specify. Again, we are drawn to a conclusion that strongly resembles Buddhism's concept of interdependence—that each part depends on all the other parts.

Emptiness: The Absence of an Intrinsic Reality

The notion of interdependence leads us directly to the idea of emptiness or "vacuity" in Buddhism, which does not mean nothingness (as often thought erroneously by Westerners) but the absence of inherent existence. Since everything is interdependent, nothing can be self-defining and exist inherently. The idea of intrinsic properties that exist in themselves and by themselves must be thrown out. Once again, quantum physics has some-thing strikingly similar to say.

According to Bohr,[7] we can no longer talk about atoms and elec-trons as being real entities with well-defined properties, such as speed and position. We must consider them as part of a world made up of po-tentialities and not of objects and facts. The very nature of matter and light becomes subject to interdependent relationships. It is no longer intrinsic but can change because of an interaction between the observer and the object under observation. Light and matter have no intrinsic reality because they have a dual nature; they appear either as waves or particles, depending on the measuring apparatus. The particle and wave aspects cannot be dissociated and complement each other. This is what Bohr called the "principle of complementarity." The phenomenon that we call a particle becomes a wave when we are not observing it. But as soon as a measurement is made, it starts looking like a particle again. To speak of a particle's intrinsic reality, or the reality it has when unobserved, would be meaningless, because we could never apprehend it.

The atom concept is merely a convenient picture that helps physicists put together diverse observations of the particle world into a coherent and logical scheme. Bohr spoke of the impossibility of going beyond the results of experiments and measurements: "In our description of nature, the purpose is not to disclose the real essence of phenomena, but only to track down, so far as possible, relations between the manifold aspects of our experience."[8] Only relationships between objects exist, not the objects themselves. Quantum mechanics has radically relativized our conception of an object by making it subordinate to a measurement or, in other words, an event. What is more, quantum fuzziness places a stringent limit on how accurately we can measure reality. There will always be a degree of uncertainty about either the position or the speed of a particle. Matter has lost its substance.

Impermanence at the Heart of Reality

In Buddhism, the concept of interdependence is also closely linked to the notion of the impermanence of phenomena. Buddhism distinguishes two types of impermanence: the gross impermanence—the changing of seasons, the erosion of mountains, the passage from youth to old age, our varying emotions—and the subtle impermanence—at each infinitesimal moment, everything that seems to exist changes. The universe is not made up of solid, distinct entities but is like a vast stream of events and dynamic currents that are all interconnected and constantly interacting.

This concept of perpetual, omnipresent change chimes with modern cosmology. Aristotle's immutable heavens and Newton's static universe are no more. Everything is moving, changing, and is impermanent, from the tiniest atom to the entire universe, including the galaxies, stars, and humankind. The universe is expanding because of the initial impulse it received from its primordial explosion. This dynamic nature is described by the equations of General Relativity. With the big bang theory, the universe has acquired a history. It has a beginning, a past, present, and future. It will die in an infernal conflagration or else an icy freeze. All of the universe's structures—planets, stars, galaxies, and galaxy clusters—are in perpetual motion and take part in an immense cosmic ballet. They rotate about their axes, orbit, fall toward or move apart from one another. They, too, have a history. They are born, reach maturity, then die. Stars have life cycles that span millions, or even billions, of years.

Impermanence also rules the atomic and subatomic world. Particles can modify their nature: a quark can change its family or "flavor," a proton can become a neutron and emit a positron and a neutrino. Matter and antimatter annihilate each other to become pure energy. The energy in the motion of a particle can be transformed into another particle, or vice versa. In other words, an object's property can become an object. Because of the quantum uncertainty of energy, the space around us is filled with an unimaginable number of "virtual" particles, with fleeting ghost-like existences. Constantly appearing and disappearing, they are a perfect illustration of impermanence with their infinitely short life cycles.

Is There an Anthropic Principle?

Despite the remarkable convergences outlined above, there is one area where Buddhism may enter into conflict with modern cosmology. This concerns the fact that the universe has had a beginning and has been fine-tuned to an extreme degree for the emergence of life and consciousness.

Copernicus's Ghost

Since the sixteenth century, the place of humanity in the universe has shrunk considerably. In 1543, Nicholas Copernicus knocked the Earth off its pedestal as the center of the universe by demoting it to the rank of just another planet revolving round the Sun. Ever since, the ghost of Copernicus has not ceased to haunt us. If our planet wasn't at the center of the universe, our ancestors thought, then the Sun must be. But it was discovered at the beginning of the twentieth century that the Sun is just a suburban star among the hundreds of billions of stars that make up our Galaxy. We now know that the Milky Way is only one of the several hundred billion galaxies in the observable universe, which has a radius of some forty-seven billion light-years. Humanity is just a grain of sand on the vast cosmic beach.

The shrinking of our place in the world led to French philosopher Blaise Pascal's cry of despair in the seventeenth century: "The eternal silence of endless space terrifies me." His anguish was echoed three centuries later by the French biologist Jacques Monod in his book *Chance and Necessity*: "Man knows at last he is alone in the unfeeling immensity of the universe, out of which he has emerged only by chance"; and by physicist Steven Weinberg, who remarked: "The more the universe seems comprehensible, the more it also seems pointless."[9]

The Anthropic Principle

I do not think that human life and consciousness arose purely by chance in an unfeeling universe. To my mind, if the universe is so large, then it evolved this way to allow us to be here. Modern cosmology has discovered that the conditions that allow for an intelligence to emerge seem to be coded into the properties of each atom, star, and galaxy in our universe and in all of the physical laws that govern it. The universe appears to have been very finely tuned in order to produce an intelligent observer capable of appreciating its organization and harmony. This statement is the basis of the anthropic principle, from the Greek *anthropos* (person).

There are two remarks to be made. First, the term "anthropic" is really inappropriate, as it implies that humanity in particular was the goal toward which the universe has evolved. In fact, anthropic arguments would apply to any form of intelligence in the universe. Second, the definition I gave above concerns only the "strong" version of the anthropic principle. There is also a "weak" version that doesn't presuppose any intention in the design of nature. It almost comes down to a tautology (the properties of the universe must be compatible with the existence of humankind), and I will not discuss it further.

What is the scientific basis of the anthropic principle? The way our universe evolved depended on two types of information: (1) its initial conditions, such as its total mass and energy content, its initial rate of expansion, and so on; and (2) about fifteen physical constants, namely, the gravitational constant, the Planck constant, the masses of the elementary particles, the speed of light, and so on. We can measure the values of these constants with extreme precision but do not have any theory to predict them. By constructing model universes with varying different initial conditions and physical constants, astrophysicists have discovered that these need to be fine-tuned to the extreme. If the physical constants and the initial conditions were just slightly different, we wouldn't be here to talk about them.

For instance, let's consider the initial density of matter in the universe. Matter has a gravitational pull that counteracts the force of expansion from the big bang and slows down the universe's rate of expansion. If the initial density had been too high, then the universe would have collapsed into itself after some relatively short time—a million years, a century, or even just a year, depending on the exact density. Such a time span would have been too short for stars to accomplish their nuclear alchemy and produce heavy elements like carbon, which are essential to life. On the other hand, if the initial density of matter had been too low, then there would not have been enough gravity for stars to form. And with no stars, no heavy elements, so no life! Everything hangs on an extremely delicate balance. The initial density of the universe had to be fixed to an accuracy of 10^{-60}. This astonishing precision is analogous to the dexterity of an archer hitting a one-centimeter-square target placed fifteen billion light-years away, near the edge of the observable universe! The precision of the fine-tuning varies, depending on the particular constant or initial condition, but in each case, just a tiny change makes the universe barren, devoid of life and consciousness.

Chance or Necessity?

How to account for that extraordinary fine-tuning? It seems to me that we are faced with two distinct choices: the tuning was the consequence of either chance or necessity (to quote the title of Monod's book). If we opt for chance, then we must postulate an infinite number of other parallel universes in addition to our own (these multiple universes form what is sometimes called a multiverse). Each of these universes will have its own combination of physical constants and initial conditions. But ours was the only universe born with just the right combination to have evolved to create life. All the others were losers and only ours is the winner. If you play the lottery an infinite number of times, then you inevitably end up winning the jackpot.

On the other hand, if we reject the hypothesis of parallel universes and adopt the hypothesis of a single universe—ours—then we must postulate the existence of a principle of creation that finely adjusted the evolution of the universe. How to decide? Science cannot help us to choose between these two options. In fact, there are several different scientific scenarios that allow for multiple universes. For example, to get around the probabilistic description of the world by quantum mechanics, the American physicist Hugh Everett has proposed that the universe splits into as many nearly identical copies of itself as there are possibilities and choices to be made.[10] Some universes would differ by only the position of one electron in one atom but others would be more radically different. Their physical constants, initial conditions, and physical laws wouldn't be the same. Another scenario is that of a cyclical universe with an infinite series of big bangs and big crunches. Whenever the universe is reborn from its ashes to begin again in a new big bang, it would start with a new combination of physical constants and initial conditions. A third possibility is the theory proposed by Andreï Linde, whereby each of the infinite number of fluctuations of the primordial quantum froth created a universe.[11] Our universe would then be just a tiny bubble in a super-universe made up of an infinite number of other bubbles. None of those universes would have intelligent life because their physical constants and laws wouldn't be suitable.

Intriguing as these notions are, I do not subscribe to the idea of multiple universes. The fact that all of these universes would be unobservable, and thus unverifiable, contradicts my view of science. Science becomes metaphysics when it is not subjected to the test of experimental proof. Furthermore, Occam's Razor bids us to cut out all the hypotheses that are not necessary. Why create an infinite number of barren universes just to produce one that is conscious of its own existence? In my work as an astronomer, I often have the good luck to travel to observatories to contemplate the cosmos. I am always awed by its organization, beauty, harmony, and unity. It is hard for me to think that all that splendor is but the product of pure chance. If we reject the idea of multiple universes and postulate the existence of just one universe, ours, then it seems to me that we must wager, just like Pascal, on the existence of a creative principle responsible for the fine-tuning of the universe. For me, this principle is not a personified god. It is rather a pantheistic principle that is omnipresent in nature, not unlike that described by Einstein and Spinoza. Einstein put it like this:

The scientist is possessed by the sense of universal causation. . . . His religious feeling takes the form of a rapturous amazement at the harmony of natural law, which reveals an intelligence of such superiority that, compared with it, all the systematic thinking and acting of human beings is an utterly insignificant reflection.[12]

He added: "I believe in Spinoza's God who reveals himself in the harmony of all that exists, but not in a God who concerns himself with the fate and actions of human beings."[13]

Buddhism Denies the Existence of a Creative Principle

The Pascalian wager I just outlined is contrary to the Buddhist approach, which denies the existence of a creative principle (or a watchmaker God). It considers that the universe doesn't need tuning for consciousness to exist. According to it, both have always coexisted, so they cannot exclude each other. Their mutual suitability and interdependence is the precondition for their coexistence. I am not totally at ease with this explanation, however.

While I admit that this might explain the fine-tuning of the universe, it seems far less clear to me that Buddhism can answer existential questions of the sort that Leibniz asked about the universe: "Why is there something, rather than nothing?"[14] I would add, "Why are the natural laws as they are and not different?" For example, it would be quite easy to imagine us living in a universe governed only by Newton's laws, but this isn't the case, for the laws of quantum mechanics and relativity describe the known universe. The Buddhist view also raises other questions. If there is no creator, the universe cannot have been created, so there is neither a beginning nor an end. The only sort of universe that would be compatible with this idea is a cyclical one, with an endless series of big bangs and big crunches.

But the scenario of the universe collapsing into itself one day in a big crunch is far from being proven scientifically. It all depends on the amount of dark matter in the universe, and this is as yet unknown. According to the latest astronomical observations, the universe does not seem to have enough dark matter to stop and reverse its expansion. They seem to indicate a flat universe that will expand forever and will stop only after an infinite time. Thus, our present state of knowledge seems to exclude the idea of a cyclical universe. As for streams of consciousness that have coexisted with the universe since the first fractions of a second after the big bang, science is still far from being able to examine this question. Some neurobiologists think that there is no need for a consciousness continuum that coexists with matter, and that the former can emerge from the latter once a certain complexity threshold has been passed.

Science and Spirituality: Two Windows into Reality

I have attempted to show that there are striking convergences between the views of reality of modern science and Buddhism. The concept of interdependence that is at the heart of Buddhism is echoed by the globality of reality implied by the EPR experiment on the subatomic and atomic scale, and by Foucault's pendulum and Mach's principle on the scale of the universe. The Buddhist concept of emptiness, the absence of intrinsic existence, finds its scientific equivalent in the dual nature of light and matter in quantum mechanics. Because a photon is a wave or a particle depending on how we observe it, it cannot be said to exist as an entity with an inherent existence. The concept of impermanence echoes the concept of evolution in cosmology. Nothing is static; everything changes, moves, and evolves, from the tiniest atom to the largest structures in the universe. The universe itself has acquired a history. I have also pointed out a potential area of disagreement: Buddhism rejects the idea of a beginning of the universe and of a creative principle that fine-tunes its properties for the emergence of life and consciousness.

The above convergences are not surprising, since both science and Buddhism use criteria of rigor and authenticity to attain the truth. Since both aim to describe reality, they must meet on common ground and not be exclusive of each other. Whereas in science, the primary methods of discovery are experimentation and theorizing based on analysis, in Buddhism, contemplation is the primary method. Both are windows that allow us to peer at reality. They are both valid in their respective domains and complement each other. Science reveals to us "conventional" knowledge; its aim is to understand the world of phenomena. Its technical applications can have a good or bad effect on our physical existence. Contemplation, however, aims to improve our inner selves so that we can improve everybody's existence.

Scientists use ever more powerful instruments to probe nature. In the contemplative approach, the only instrument is the mind. The contemplative observes how his thoughts are bound together and how they bind him. He examines the mechanisms of happiness and suffering and tries to discover the mental processes that increase his inner peace and make him more open to others in order to develop them, as well as those processes that have a destructive effect in order to eliminate them. Science provides us with information but brings about no spiritual growth or transformation. By contrast, the spiritual or contemplative approach must lead to a profound personal transformation in the way we perceive the world and act on it. The Buddhist, by realizing that objects have no intrinsic existence, lessens his attachment to them, which diminishes his suffering. The scientist, with the same realization, is content to consider the acquired knowledge as an intellectual advance that can be used to further his work without changing fundamentally his basic vision of the world and how he leads his life. When faced with ethical or moral problems that, as in genetics, are becoming ever more pressing, science needs the help of spirituality in order not to forget our humanity. As Einstein said so well:

The religion of the future will be a cosmic religion. It will have to transcend a personal God and avoid dogma and theology. Encompassing both the natural and the spiritual, it will have to be based on a religious sense arising from the experience of all things, natural and spiritual, considered as a meaningful unity. . . . Buddhism answers this description. . . . If there is any religion that could respond to the needs of modern science, it would be Buddhism.[15]

NOTES

1. Matthieu Ricard and Trinh Xuan Thuan, *The Quantum and the Lotus* (New York: Crown, 2001; New York: Three Rivers Press, 2002).

2. Werner Heisenberg, "Wolfgang Pauli's Philosophical Outlook", chapter 3 in *Across the Frontiers* (New York: Harper & Row, 1974).

3. Albert Einstein, Boris Podolsky, and Nathan Rosen, "Can quantum-mechanical description of physical reality be considered complete?" *Physical Review* 47 (1935): 777–80.

4. John S. Bell, *Speakable and Unspeakable in Quantum Mechanics*, 2nd ed. (Cambridge: Cambridge University Press, 2004); A. Aspect, P. Grangier, and G. Roger, *Phys. Rev. Lett.* 49 (1982): 91.

5. W. Tittel, J. Brendel, H. Zbinden and N. Gisin, "Violation of Bell inequalities by photons more than 10 km apart," *Phys. Rev. Lett.* 81 (1988): 3563–566.

6. Ernst Mach, *The Science of Mechanics: A Critical and Historical Exposition of its Principles* (Chicago: Open Court, 1893).

7. Niels Bohr, *Atomic Theory and the Description of Nature* (Woodbridge, CT: Ox Bow Press, 1987).

8. Ibid., 18.

9. Blaise Pascal, *Pensées*, trans. A. J. Krailsheimer (New York: Penguin Putnam, 1995); Jacques Monod, *Chance and Necessity* (New York: Knopf, 1971); Steven Weinberg, *The First Three Minutes* (New York: Basic Books, 1977).

10. Hugh Everett III, "On the Foundation of Quantum Mechanics" (Ph.D. thesis, Princeton University, 1957).

11. Andreï Linde, "The self-reproducing inflationary universe," *Scientific American* 271, no. 5 (November 1994): 48–55.

12. In Alice Calaprice, ed., *The Quotable Einstein* (Princeton: Princeton University Press, 1996), 151.

13. Ibid., 147.

14. Gottfried Leibniz, *Discourse on Metaphysics* (Chicago: Open Court, 1976).

15. Quoted by Thinley Norbu in "Welcoming Flowers" from *Across the Cleansed Threshold of Hope: An answer to the Pope's criticism of Buddhism* (New York: Jewel Publishing House, 1997).

PART V

Personal and Scientific Search

ৡৣ

Ordinary Faith, Ordinary Science

WILLIAM D. PHILLIPS

Introduction

I was born into a family that took religion seriously. We said grace before meals; we said prayers before bedtime; we belonged to a Methodist church and attended church school and worship services almost every Sunday. In short, we were like most of the other families I knew as a child. It never occurred to me that religious faith was anything but a natural, ordinary part of life.[1]

As early as I can remember, I was interested in science. At first, I suppose, this was simply the usual childlike curiosity about the way things work. As I became aware that this kind of curiosity could be a profession, I understood that this was what I wanted. By about the age of ten, I knew I wanted to be physicist. Maybe it was because physics was cleaner and less smelly than chemistry and biology (although I am still fascinated by those subjects) or maybe it was because physics addresses the most fundamental questions about the way the universe works (although at that time I had almost no notion of what those questions were). For whatever reasons, I was to be a physicist, and so I am.

It didn't seem to me that there was a fundamental conflict between my interest in science and my grounding in religion. Of course, I knew that the stories in the Bible, especially accounts of the creation, were in literal conflict with the scientific understanding of the origins of the universe and its inhabitants. But by the time I could see those conflicts clearly, I had also learned about the variety of literary expression, and the ways in which deep meaning emerges from devices like metaphor, allegory, and poetry. My parents at home and my pastors at church encouraged me to pay attention to the spiritual message of the scriptures. Science was one thing and religion was another, and there was no problem.

I am still a church member; I'm part of the congregation of Fairhaven United Methodist Church in Darnestown, Maryland, and I sing in the gospel choir. Our family says grace before meals, goes to church almost every Sunday, and I pray less often than I should. I am more liberal in my religious views than some in my church and more conservative than others. In short, my religious life is pretty conventional. So is my scientific life. I am a member of the American Physical Society, the professional organization for physicists. I write papers and give talks that are greeted with the same healthy mixture of respect and skepticism that any physicist should expect. I am privileged to lead a group of some fifteen to twenty scientists at the National Institute of Standards and Technology, people whose enthusiasm and intelligence make it a joy to come to work each day. As a professor at the University of Maryland, I have the pleasure of teaching students whose questions provide challenge and satisfaction. In short, I am an ordinary physicist.

Being an ordinary scientist and an ordinary Christian seems perfectly natural to me. It is also perfectly natural for the many scientists I know who are also people of deep religious faith. For others, however, it appears strange, even astonishing, that someone could be serious about science and also serious about faith. Here, I will try to give an impression of how these two aspects of my life work for me, and how they influence and inform each other. Mostly, this is the witness of an ordinary person that one can be serious about science and serious about faith.

My Science

I think of myself as a quantum mechanic. That is, just as a car mechanic works on automobiles in a practical way, I work on the quantum nature of atoms and light in a practical way. Quantum mechanics is the theory of physics that describes how things work at the submicroscopic level of atoms and photons (particles of light). It is a marvelously successful theory, and, as far as we can tell, it correctly describes all of the ordinary phenomena we experience in everyday life in addition to a wealth of phenomena that are seen only with the specialized instruments of quantum physicists.

The behavior of things at the microscopic, quantum level is much different from the familiar behavior of larger, macroscopic objects. For example, in ordinary life we are used to saying that you can't be in two places at the same time. In the quantum world, an atom, a photon, or an electron being in two places at the same time is commonplace. In the macroscopic world of ordinary experience, objects have properties independently of whether we look at those properties or not. A one-way sign on a certain street points either east or west. We may have to look at it to know which way it points, but, if it is a conventional sign, we don't have to look at it to know that it *does* point one way or the other. In quantum physics, an atom can be in a superposition of pointing east and pointing west (that is, it is pointing both east and west). And we can show by experiment that it would be wrong to suppose that it is pointing one way or the other before we look.

If these features of quantum mechanics seem strange and confusing to the nonphysicist, be assured that they are equally confusing to physicists. We don't claim to understand why things work in this odd way; we just know that they do, and we do useful things with that knowledge. Many of the things we take for granted in modern life (consumer electronics, for example) only exist because scientists and engineers have understood the peculiar aspects of quantum physics and made devices that rely on them.

I work with these strange behaviors on a daily basis. To me they are as familiar as the workings of an internal combustion engine are to a car mechanic. If I stop to think about how weird quantum mechanics is, I am as confused as anyone. But I can use my knowledge of that weird behavior to achieve results that are as reliable as the operation of an automobile. (Of course, automobiles are not perfectly reliable, and neither are my laboratory experiments, but these failures are not the result of fundamental problems with understanding the mechanics of cars or the quantum mechanics of atoms.)

My specific research has been in the laser cooling and trapping of atoms.[2] Rather surprisingly, one can cool a gas of atoms by shining light on it. The temperatures achieved are among the lowest ever seen for any substance—less than a millionth of a degree above absolute zero. Such low temperatures mean that the atoms are moving extremely slowly, less than a centimeter per second. (This is to be compared with hundreds of meters per second for atomic gases near room temperature.) When atoms move this slowly, their wave-like character becomes increasingly evident. Quantum mechanics tells us that all particles also behave like waves, another strange and wonderful aspect of nature. When the particles are heavy or moving rapidly, the wavelength is so small that the wave nature is not usually evident. But, when the velocity of something as light as an atom is reduced below a centimeter per second, the wavelength can become longer than the wavelength of visible light. Then, the wave nature of the atom may become evident even on the macroscopic level, at distance scales much larger than atomic dimensions.

In some of our experiments, we put gas atoms into a special state called a Bose-Einstein condensate. Only in the past several years has it been possible to make such a state.[3] In a condensate the atoms can have a wavelength larger than a tenth of a millimeter: large enough for someone with good eyesight to see with the naked eye. In a sense, my research group and others like it are bringing some of the strange aspects of quantum mechanics from the subatomic to the macroscopic world. Although our intuition about what happens under these circumstances is often not very good, we have always found that quantum mechanics continues to give a correct description of what we observe.

Experiments on laser-cooled atoms and Bose-Einstein condensates have both fundamental and practical applications. On the practical side, as of the fall of 2001, at least three countries are using atomic clocks with laser-cooled atoms to provide their national time standards. The future promises even more excitement. We hope to use laser-cooled atoms as *qubits* (quantum bits) in quantum information processors—quantum computers that will be different from present computers in a more fundamental way than today's machines are from the abacus. These new quantum computers would have at their heart the quantum weirdness that is so intriguing to physicists, and might be able to solve problems that are inaccessible to ordinary computers.

My Faith

I am uncomfortable being described as religious. I suppose that for me the term conjures up an image of someone overly concerned with the outward appearances of religious practice rather than with the spiritual core of religion. So, I have often preferred to describe myself as a person of faith.

The author of Hebrews describes faith as "... the substance of things hoped for, the evidence of things not seen" (Hebrews 11:1 KJV). I find this declaration both beautiful and profound. The juxtaposition of the solid words "substance" and "evidence" with the ethereal descriptors "hoped for" and "not seen" emphasizes that faith is belief that has a different foundation from that associated with scientific understanding.

One of the participants in the Physics and Cosmology Panel of Science and the Spiritual Quest II posed this question: "Can you imagine any evidence that would make you stop believing in God?" The question has great importance because any scientific hypothesis must be falsifiable. That is, one must be able to specify what would show the hypothesis to be false. Statements that are not falsifiable are not scientific statements. My answer to the question about God is: "No, there is nothing that would cause me to stop believing in God."[4] By my definition, this means the belief is not a scientific one.

That said, I nevertheless emphasize that my scientific understanding supports my faith. My faith may be nonscientific (I don't say "unscientific"), but it is not irrational! When I examine the orderliness, understandability, and beauty of the universe, I am led to the conclusion that a higher intelligence designed what I see. My scientific appreciation of the coherence, and the delightful simplicity of physics strengthens my belief in God.[5] The structure of the universe seems uncannily suited to the development of life. Small changes in any number of fundamental constants of nature (those numbers that describe, for example, the force between two electrons) or initial conditions for the universe (like the total amount of matter) would have made it impossible for life as we know it to develop. Why is the universe so finely tuned for the existence of life? More to the point, why is the universe so finely tuned for the existence of us? A simple answer is that had it not been so we wouldn't be here to ask the question (the anthropic principle). This leaves unanswered the question of why, out of all the essentially infinite possible universes that could have been, the one that does exist, supports intelligent life. It seems so improbable that many conclude an intelligent creator must have designed the universe this way.

Does this constitute legitimate scientific evidence for an intelligent creator? It may. But it is not universally compelling. Better and more intelligent scientists than I, people who are better acquainted with the order and beauty of the cosmos, have reached the opposite conclusion. (And better scientists have reached the same conclusion as I have.) Hypotheses about multiple universes address the issue of the incredibly low probability of having a universe suitable for life. (Although these hypotheses, for the moment at least, have no more hard evidence to support them than does a belief in God.)

I have a feeling (a feeling without much scientific or theological support) that we will never find truly convincing scientific evidence about the existence of God. I suspect that God does not leave his "fingerprints" on his handiwork.[6] One sage noted that if there were convincing evidence of God's existence, what then would be the use of faith? Nevertheless, many scientists find the scientific evidence to be compelling enough that they believe in an intelligent creator who set into being and into motion all that we see around us. Many subscribe to a belief in what is sometimes called "Einstein's God,"[7] an embodiment of the intelligence and order behind creation but not a personality who cares about

and interacts with the creation. In other words, not the God of traditional religion. The variety of such belief is as great as that of the believers. In a particularly beautiful and articulate expression of this kind of
belief,[8] one scientist identifies her favorite hymn: "Immortal, Invisible,
God Only Wise" by Walter Chalmers Smith (1867):

> Immortal, invisible, God only wise,
> In light inaccessible hid from our eyes,
> All laud we would render: O help us to see
> 'Tis only the splendor of light hideth thee.

"Immortal, Invisible" is a great hymn, but with the distance at which it
puts God, it doesn't even make my top twenty list. Among *my* all-time
favorites is "In the Garden" by C. Austin Miles (1913), with its sweet re-
frain:

> And he walks with me, and he talks with me,
> And he tells me I am his own;
> And the joy we share as we tarry there,
> None other has ever known.

"In the Garden" expresses my belief in a personal God, a God who is
both the creator of the universe and is intimately concerned with the
welfare of the creatures of that universe. "Einstein's God" is not nearly
enough for me. I believe in a God who wants good things for us, and
who wants and expects us to care for our fellow creatures. I believe that
God wants genuine, loving relationships with us, and wants us to have
such relationships with each other. I don't see how I can call upon the
beauty and symmetry of nature or the astronomically improbable fine-
tuning of the universe in support of this kind of belief. So why do I believe in a personal and loving God?

Another favorite hymn comes to mind, perhaps the first hymn I
learned as a child: "Jesus loves me! This is I know, for the Bible tells me
so" (Anna B. Warner, 1860). I believe in the loving nature of God because of what I have been taught from the scriptures, because of the traditions handed down from ancient times, and because of the wisdom received from my parents and teachers. But there is more. I am convinced

of the truth of what I believe about God because I can feel God's presence in my life and in the world. Prayer comforts me and helps me to make good choices. People are kind and good, sacrificing their own welfare for the welfare of others. All of this is part of the "evidence of things not seen" that convinces me of the reality of a loving God. Of course I am well aware of all the arguments in the other direction: secular meditation has all the benefits of prayer; psychological and/or survival value lead to altruistic behavior. Nevertheless, I believe.

Are these beliefs held without any doubts? Hardly! I have repeatedly asked myself whether this belief in God is just a psychological crutch or an unreflective acceptance of tradition. I wonder from time to time whether there might be a God, but I've just got it all wrong: that God doesn't care or doesn't exist as a true personality, but only as some ill-defined sum of the myriad consciousnesses of the universe. I don't have those kinds of doubts about physics, and that is an important difference between my science and my faith. But I accept that such doubts are part of a life of faith. The story of Thomas (John 20:24–29 RSV) is, I believe, part of our scripture to comfort us in our doubts. If Thomas, who was a disciple and a daily companion of Jesus, had doubts, then it's not so bad when we do as well.

Among the things that stir up doubt in a reflective person of faith are the difficult issues faced by anyone, scientist or not, who claims to believe in a personal, loving, active God. Foremost in my mind is "Why is there suffering in the world?" Of course, some suffering is the result of the sinful acts of those who suffer. People who abuse drugs and alcohol suffer as a result. Less easy to accept is that innocent people suffer because of the misdeeds of others: the children and relatives of the drug abusers, for example. But if God wants to have genuine relationships with us, then we must be free to reject God and all that God wants for us. Suffering of both the guilty and the innocent, as the result of sin, would seem to be an unavoidable by-product of God's gift of free will to us. Perhaps most difficult to understand is the suffering of innocent people because of random events beyond human control. Why did God create a world in which volcanoes destroy cities, or disease brings unspeakable pain to small children? I simply don't know! This question is as old as religion itself, and as puzzling today as ever. As I see it, the book of Job was written to address the question, and my understanding of the answer given there is that there are simply some things that we are not going to understand. It may be that to have a world in which God's creatures are truly free to make choices, God had to allow the possibility of such undeserved suffering. It may be so, but I certainly don't know.

Another difficult problem, especially for Christians, is the status of those of other faiths. As a Christian I believe that Jesus reveals God. I believe that Jesus is the living proof of God's desire for person-to-person relationships with us. Jesus, through his sacrificial life and death, reconciles us to God and guarantees us eternal life. So what about everyone who does not accept this view of Jesus? What about all those who accept the principles of behavior preached by Jesus, and who live up to those principles far better than I? After all, Jesus often said that he was only preaching what the Law and the Prophets had taught long ago. Again, I don't know! For me, Jesus is "the way, and the truth and the life" (John 14:6 RSV). But I cannot claim to speak for God and to say that others who are on a different spiritual journey are taking the wrong path. A great blessing of participating in SSQ II has been what I have learned from other scientists of different faiths. I have been far more impressed by the similarities of our spiritual experience than by the differences. I certainly won't claim that all religions are the same, but when so many have such common features, I find it hard to argue that the loving and personal God I experience is not at work in the hearts of those people of other faiths.

I believe that I and other people of faith understand some important things about God and God's purposes, but I also believe there is much that we do not and will not understand, at least in this earthly life. As Saint Paul put it: "For now we see in a mirror dimly, but then, face to face. Now I know in part; then I shall understand fully, even as I have been fully understood" (1 Corinthians 13:12 RSV). From such a position of ignorance, I feel that I must be cautious about being dogmatic in my beliefs, and I must remain open to the insights that others might bring.

How It All Fits Together

I have said that belief based on faith is different from belief based on scientific evidence. Why do I believe that there are these two ways of knowing things? As a scientist trained to accept only reliable, reproducible evidence in support of hypotheses, why do I believe in "the evidence of things not seen"? Why not?! I think even those scientists who most firmly believe that only empirical evidence leads to truth find room in their lives for love and romance. Even if they believe that love is just biochemistry, I doubt very much that, in a tender, romantic moment, they behave that way. If we are all comfortable in surrendering an important part of our lives to something as clearly apart from scientific rationality as love, then why not faith? I am not arguing that one should believe in God because science cannot explain love. I am arguing that even if science *could* explain love, it is self-evident that there would be great value in continuing to see and embrace love in a nonscientific way, and that most of us will continue to do so. If we are willing, even eager, to do that, why should we be less willing to embrace faith?

I realize that this argument is a bit flippant, but I still think it has merit. There is no reason to believe there is one and only one way to look at life. I am very much drawn to the observation of physicist Freeman Dyson, who says that science and religion look at the same reality through different windows.[9] It seems to me that life would be rather dull if we only looked at it through the window of science.

Another useful insight, well explained by Howard Van Till, is that science and religion address different kinds of questions about reality.[10] Science can address questions about how things work and what sequence of events led to the present circumstances; religion can address questions about our relationship with God and how we should behave toward others. Trouble comes when we address questions to the wrong discipline. I see the book of Genesis telling us about God as magnificent creator (chap. 1) and as a personal, involved parent (chap. 2). Cosmology tells us about stellar evolution and biology tells us about the origin of species. Trying to learn about cosmology from Genesis not only poses the question to the wrong discipline, it runs the risk of missing the important spiritual messages contained in Genesis.

These descriptions of the relation of science and religion might seem to indicate that they are completely separate disciplines, using completely different methods to address completely different problems. I don't see things that way. As a Methodist, I was taught that belief is founded on the four pillars of Scripture, Tradition, Reason, and Experience (the "Methodist Quadrilateral"). I see strong parallels between these and the foundations of scientific knowledge. Scripture (the Bible) and tradition (the wisdom of religious thinkers throughout history) represent received knowledge. Science has plenty of that. We read the classic texts in physics, and we generally accept the descriptions of experimental evidence without repeating the experiments ourselves. In that sense, we accept a lot of science on faith. There is a key difference, however: in science we could in principle verify the described experiments at any time, and we have a multitude of modern witnesses who have contributed their own verifications. Such verification is not in general available for the received knowledge of religion.

I see reason and experience as being even more similar in science and religion. There is a common misconception that religion must ignore reason and experience in favor of received knowledge, but that is not at all consistent with my religious tradition. Religious thinkers at least as far back as Saint Augustine have taught that when clear empirical evidence contradicts the scriptures, we are misinterpreting scripture. So, if the methods of science and religion are not so very different, and they look at the same reality through different windows, can science and religion in fact work together? Certainly when moral and ethical questions require scientific knowledge, it seems natural, even imperative that they do so. If, for example, we want to determine the advisability of distributing genetically modified foods and grains in impoverished countries, we need to understand both the science and the ethics. That sort of cooperation, where each brings something different to the same problem, seems obviously worthwhile.

Scientific discoveries can also provide support for historic religious teaching. Take, for example, the teaching of many religious traditions that all of us are brothers and sisters in the parenthood of God. Modern biology confirms the genetic identity and common ancestry of all people. Continuing instances of inhumanity to others, even within nuclear families, gives scant hope that such scientific knowledge will dramatically alter behavior, but it certainly confirms traditional teaching. I also believe that science and faith intersect because God wants us to discover as much as possible about the universe he created. Just as good parents want their children to learn as much as they can on their own, I believe that God rejoices with us in each new discovery. I believe that God wants us to enjoy abundant life through all the opportunities he gives us, including scientific discovery. And I believe God calls us to make the world a better place by increasing our knowledge of it. I believe that scientific research is a deeply religious calling. It is one of the ways in which God makes us partners in a continuing creation.

But this is all simply an expression of my religious belief about the value of science. What about something linking religious belief and scientific knowledge more directly? Studies of the fine-tuning of the universe and the anthropic principle, along with examination of hypotheses about multiple universes and about intrinsic constraints on physical laws and constants, may someday give far more convincing evidence of intelligence behind creation. (Or they may not.)

Another place where scientific investigation might make significant contributions to religious belief is the area of human consciousness. I find the fact of human consciousness and free will to be a strong argument for some sort of transcendence. If we truly have a free will, if our actions represent true choice and not just results of biochemical reactions following deterministic or random processes, then where does that will come from? If there is only physics and chemistry, where does decision come from? Of course, it may be that our impression that we have free will is illusory, or it may be that free will emerges from a sufficiently complex system, all of whose components are deterministic or random. But I find these possibilities unconvincing and find it simpler to believe in a transcendence that provides something beyond determinism or chance. I call that transcendence God. But, considering the poor state of our scientific understanding of human consciousness and free will, my conclusion about the necessity of transcendence is not particularly well-founded. A better understanding of consciousness, which may come from future scientific investigation, could significantly change this situation.

Could science prove God? Let us imagine for a moment that we find convincing evidence that there are no intrinsic constraints on how the universe might have been constructed (that is, any combinations of fundamental constants and initial conditions were allowed). Imagine that we find powerful arguments against the multiple universe pictures. And let us also imagine that further investigation solidly confirms that extremely tiny deviations from the actual conditions of our universe would have resulted in an uninteresting wasteland with no stars or planets, let alone intelligent life. Such a situation might well lead most reasonable people to believe that the hypothesis of an intelligent creator is far simpler than the hypothesis of an undirected, spontaneous, and naturalistic birth of the universe. In other words, it might turn out that belief in God becomes by far the most reasonable *scientific* conclusion.

This would be, for me and for many, a very satisfying outcome (although I strongly doubt that it will come to pass—I doubt that God has left such clear "fingerprints"). But, it would represent scientific support for only a part of my belief in God, and a small part at that. The scenario I have described would not touch on the personal, loving God whom I know.

Can I imagine that science could support my belief in a personal God in the same way that I have imagined it might provide increased, convincing support for the concept of an intelligent creator? I doubt it! Let us suppose that we wanted to test whether God is active, loving, and caring. We set up a controlled experiment to test the efficacy of intercessory prayer. (Such experiments have in fact been done, so far with inconclusive results.) Let us assume we find that indeed those for whom we have been randomly assigned to pray are healed at a significantly higher rate than those for whom we do not pray (even though the patients do not know whether or not they are being prayed for). Do we conclude that God is loving and kind because he exerts his healing power, or that he is fickle and shallow because he responds to suffering according to the arbitrary, random choices of the investigators?

The difficulty of such a question mirrors a continuing theological dilemma I have about intercessory prayer: I find it difficult to understand why my prayers for a suffering friend would induce God to exert healing power when I believe that God already loves my friend far more deeply than I do. Yet I pray. I don't see that experiments can resolve this question, or provide convincing evidence of a personal God.

All of this discussion about whether we might test for God's action in the world raises the question of how a God, caring and active in our world, accomplishes action within the framework of physical laws that have always been seen to be trustworthy descriptions of how God's universe works. Van Till cites the unwavering validity of physical law as evidence of God's faithfulness to creation.[11] Then what of miracles or violations of physical law as reported in the scriptures or in more recent religious experience? I have a number of observations. First, we should recognize that the writers of the scriptures did not have the same view of the immutability of physical law as we do today. What we would call "magic" was seen as an everyday occurrence, and accepted as a part of life, so the spiritual message delivered by the account of a miracle did not likely include the idea that God sometimes suspends otherwise immutable physical law. This is not to say that I know God could not or would not do that, or that we would be able to verify such suspensions if they did occur (science, after all, is mostly about reproducible phenomena; irreproducible phenomena are generally discarded as resulting from untrustworthy observations). On the other hand, it could be that God's interventions are more subtle, occurring at the level of quantum probability, where physics allows a multiplicity of more or less probable outcomes, from which God might choose without any apparent contradiction of physical law.

These are all interesting and entertaining questions. Nevertheless, I believe they are far less important than questions of how we, as God's creatures, should act toward our fellow creatures. When I was a boy, I was fond of the story of Samuel (1 Samuel 3:2–10). The boy Samuel hears God calling him in the night, and believes it to be his mentor, Eli. Eli sends the boy back to bed, but when this happens again and still a third time, Eli perceives that it is God calling Samuel, and Eli advises Samuel how to respond. When, as a boy, I would hear soft sounds in the night, and would imagine that I heard my name, I thought that perhaps it was God calling me. As I grew, I realized that night sounds play tricks on the mind and there was nothing of substance in my imaginings. Now, I know that I had it pretty much right the first time. God is calling me, and each of us, all the time to do the work that needs to be done. I am reminded of another of my favorite hymns, "Here I Am, Lord" by Dan Schutte (1981) (third verse and chorus):

> I, the Lord of wind and flame,
> I will tend the poor and lame,
> I will set a feast for them,
> My hand will save.
> Finest bread I will provide,
> till their hearts be satisfied,
> I will give my life for them
> Whom shall I send?
>
> Here I am Lord.
> Is it I Lord?
> I have heard you calling in the night.
> I will go Lord, if you lead me.
> I will hold your people in my heart.[12]

One of my favorite passages of scripture is Matthew 25:31–46. It is not a favorite because I find any comfort in it, but because it seems to tell me most clearly what God expects of me. Here Jesus makes it clear that how we treat those who are hungry, ill, and oppressed is of extraordinary importance to him. He tells us, "as you did it to the least of these . . . you did it to me" (Matthew 25:40 sv). This responsibility to help those who are in need is awesome and daunting. There is a lot to be done. We should probably get on with it.

NOTES

1. I thank all of the members of the Physics and Cosmology Panel of Science and the Spiritual Quest II, and to the staff who made possible our meetings. Our discussions, and the ways in which they helped to focus my thinking about these issues, were crucial in what I have written here. The relationships formed in those meetings have been a special blessing. I am also deeply grateful to all the many people who have shaped my faith and my science over the years: my parents, my pastors, my teachers and mentors, my friends and colleagues, the members of the Sunday school and Bible study classes I have enjoyed over the years, and of course, my family who have always been so supportive. Finally, I thank God for all the love, beauty, and wonder in this Creation.

2. W. D. Phillips, "Laser cooling and trapping of neutral atoms," *Rev. Mod. Phys.* 70 (1998): 721–41.

3. M. H. Anderson, J. R. Ensher, M. R. Matthews, C. E. Wieman, E. A. Cornell, "Observation of Bose-Einstein Condensation in a Dilute Atomic Vapor Below 200 Nanokelvin," *Science* 269 (1995): 198.

4. Of course, the honest answer is that I don't know if there is something that would make me stop believing in God. Others with stronger faith than mine have had it shattered by personal or global tragedies. I hope that would not happen to me, but I don't know for sure.

5. I sometimes wonder if the reason physicists are more likely to be believers than are biologists is that physicists see a simpler, cleaner, more orderly and understandable world than do biologists.

6. Using personal pronouns like "he" and "his" does not mean that I believe God is male. Rather, it means that I believe God is personal. I believe that the Bible contains appropriate male and female images of God, and that no single image, or even any set of images can give us a complete picture of God. I believe that God is our mother, father, sister, brother, friend, and much more.

7. There has been considerable discussion about just what was Einstein's view of God, since he sometimes used rather personal references to God, and at other times insisted on a rather impersonal view of God.

8. Ursula Goodenough, *The Sacred Depths of Nature* (New York: Oxford University Press, 1998), 13.

9. Freeman Dyson, Acceptance Address upon receiving the Templeton Prize for Progress in Religion, Washington National Cathedral, Washington D.C., May 16, 2000.

10. Howard J. Van Till, *The Fourth Day: What the Bible and the Heavens Are Telling Us about Creation* (Grand Rapids, MI:Eerdmans, 1986).

11. Ibid.

12. Text and music, copyright 1981, OCP Pulblications, 5536 NE Hassalo, Portland, OR 97213. All rights reserved. Used with permission.

13

The Other Outlook

KHALIL CHAMCHAM

Introduction

A metaphorical outlook is richer than its physiological counterpart, as it extends over a much broader field of representation and conceptualization than the mere act of direct observation. Our understanding of the universe and its constituent parts (galaxies), the representations we use to describe it, and the idea—often prejudiced —that we have of our fellow man, are directly linked to the way in which we view the outside world in contrast to the way in which we look within ourselves. Our view of the other—our fellow man—is an extension of the view we have of ourselves: for better or worse, it is that part of ourselves that needs the other to express itself. In our social interactions, to be observed comforts us in our sense of being and to observe is to assert our place in the world. In this way, at our individual level, there is no absolute "reality." All reality is the product of what the world imparts to us and what we can view (or see).

Our place in the world and the rise of civilization are not merely determined by our physical existence here and now. They are also the result of the way we look inward and consider the road ahead. Likewise, the survival of our culture and the emergence and progression of our consciousness cannot occur without a vision and a force that tend toward that which we are not—and that feed our aspirations. It is this force that prevents us from sinking into a state of vegetation and ensures that our history is not limited to only the story of our biology. History draws its resources from the future, not the past.

In these times of religious and inter communitarian crises, marked by our failure to look one another in the eye, it is precisely by looking to the other that new horizons may open up and crises be defused. Yet I cannot hope to make steps toward the other without changing the way in which I view him or her or myself. This will require a predisposition to relinquish part of my narcissism in favor of the other, and recognition of the feelings of rejection and violence that I harbor toward him. This also implies the potential of a rebirth of oneself and the capacity to experience love instead of the confusion between our psychology and our social identity that leads to our contempt for the other.

The quest for one's self is the very opposite of a retreat within one's self and one's cultural and religious values—even if some societies and individuals see retreat as a sign of the strength of their values, never mind if this implies a slide towards decadence and closure from progressive movements. This tends to be accompanied with distrustful and malicious looks, as the meeting of eyes can expose the risk of transparency and lead to situations beyond our control.

A quest of this kind cannot be one-dimensional, as it is an invitation to encounter the world and the prospects it has to offer. It is a unique experience fortified by the multiple dimensions that open themselves to the fields of knowledge and sensibility. Whether through art, religion, science, or meditation, the driving principle is the harmony of life—being—and any experienced or desired dichotomy merely reflects our need to reject any experience of difference and conjecture. Neither is the quest for self a narcissistic exercise: it is the exposure of one's self to the forces of the world and a commitment to use these forces in the service of others and life.

In the pages that follow, I will make an appeal for inquiry without seeking to provide answers or explanations. Indeed, I do not think a theory exists, let alone a doctrine, for this quest: merely the sensitivity of observations and the love of knowledge and novelty. There are, for those who are interested, countless bibliographical references, whether in the fields of astrophysics, theology, or philosophy, to satisfy each and every one's curiosity and particular questions. My purpose here, however, is to explore border areas, not the urban areas where everything has already been mapped.

Scientific Quest

One of the peculiarities of our universe is that it allows itself—its stars and galaxies —to be discovered in the eye of the beholder, even if only in part. Our observational horizon being limited, we will never know what portion of the universe will remain forever beyond our grasp. The universe has also generated the necessary conditions for the emergence of consciousness and the development of language to interpret and codify these laws. Whether this was programmed from the start or whether the laws governing evolution and matter are the result of pure chance is beyond the power of science to answer and is better left to theologians and philosophers to address.

The appearance of humanity is one of the most recent events in the history of the universe and, to our knowledge, we are the only species that has been able to identify the constituent elements and study the laws that govern them (though this does not preclude the existence of other species elsewhere in the universe that may be the result of other forms of evolution). We are the newly born, yet, paradoxically, it is at this point, at the tail end of the chain, that we have been able to retrace the history of the universe to its origin, the big bang. From the study of atoms in interstellar space to the physiology of the gaze to the consciousness that nourishes it, on one hand it seems difficult to retrace the chain of causality and, on the other, evidence of a possible "program" remains elusive.

Galaxies are the bricks of the universe and the study of their formation and evolution is one of the major keys to our understanding of the structure and evolution of the universe.[1] All the light of the universe comes to us from the stars, the celestial bodies that contain the answers (directly or indirectly) to all the big questions concerning dark matter, the origin of life, or the existence of other civilizations.

Put simply, the history of stars is regulated by a delicate balancing act between the gravitational force that tends to compress the primordial gas from which they were formed, and their internal thermal pressure, which tends to counteract the force of gravitation. These thermal forces are the result of thermonuclear reactions that are themselves the result of the density and extremely high temperatures that are found in the core of the star.

The stars that are visible to the naked eye are in close proximity to the solar system and only represent a tiny fraction of the stars that make up the Milky Way. For the most part, they are similar to the Sun, with a life expectancy of approximately ten billion years. The larger the star, the shorter its life expectancy, due to the gravitational forces that increase in relation to the mass of the star to the point where accelerating nuclear combustion breaks the equilibrium with gravity. The star then explodes and its matter is scattered through the interstellar space from which it was originally formed. A new generation of stars will be born from this matter in a new cycle of equilibrium.

Another remarkable effect is that the birth of stars injects enough energy into the surrounding interstellar gas to restabilize it and stop the formation of stars. Hence, the process is self-regulating and intermittent, each phase repeating itself, increasing in complexity at each level. It is this intermittence and the accompanying complexity that will generate the chemical elements that are necessary for the emergence of life and the necessary conditions for evolution to occur.

What we witness with the formation of stars is a cosmic phenomenon of self-regulated intermittence as a result of which the object that is produced gives rise to the causes that will hinder its own progression, while the absence of the object will generate the conditions of its emergence. It would be more accurate to talk about the process underlying the object as opposed to the object itself, but both are linked by a chaotic causality that allows for emergence to be produced along a line of stability. I note in passing that the most remarkable cosmic phenomena (including the universe itself) evolve along a "stability line" whose physical conditions of realization are barely negotiable.

Like all cosmic phenomena, we only have access to the phenomenon of the formation of stars through the intermediary of our perspective (direct or with the help of instruments), on the scale of a human life whose timescale barely corresponds to a fraction of one-millionth of the life expectancy of the process itself. The information taken in by our perspective can only be drawn from the past because of the vast distances and the finite speed at which light travels, and we neither have access to the past nor to the future of what we observe. Everything happens, however, as if our physical absence from the world during billions of years of evolution before our own emergence as a species has been made up for by our consciousness, expressed through all the different knowledge processes. At the moment in which we observe the universe, its story opens itself to us. Hence, our sense of being in the world changes while increasing our knowledge capital. Our new vision of the world and the idea of its origin is altered. From poetic contemplation to scientific rationalization, it is the language of mathematics that opens the way to new horizons, future and past, which neither common sense nor direct vision or experimentation can decipher and help us to understand.

What is it then that connects our brain to the world in such a way that it can express it in the form of equations and explain its laws? Does the world reveal itself to the intention or the interest that is expressed in its regard, or is our brain merely activating information that has already been secreted to it? Either way, there is a two-way perspective, and the link between the observed-subject and the observer-subject is not a passive but a dynamic link that we still need to understand, though some of the keys of this understanding are being revealed by quantum mechanics and cosmology.

Besides, what is it that dictates the capacity of our brain to reproduce cosmic history—and, for that matter, any history or the history of the microphysical world—and which needs the capacity to go back in time? Is it the nature of time itself that still eludes us, or is it the existence of a "capital" of cosmic information that is accessible from any point in the universe and whose heritage we carry? This heritage would be assumed at the level of being, that is, by our "capacity to be" and being "aware of one's own being." These questions deserve to be developed in detail. However, I ask them as a preamble to the following reflections on the question of the organization of "matter" from the most simple inert matter such as a hydrogen atom to the most complex form that is reflexive and capable of analyzing its environment (such as a virus).

Quest for Origins

In the past, others observed the universe and formed their own opinions on the sky and what constitutes it. What was the nature of our ancestors' perspective? What information did they derive from it? Did they transform it into an additional intelligence as homo sapiens transforms information into knowledge? Did they feel called upon to observe the world? If all information materializes in the brain, can we suppose that the brains of our ancestors materialized their observations as knowledge or internal experience as homo sapiens endeavors to do? As a result, was their information/knowledge capital transformed and future evolution modified?

The expansion of the universe is an ongoing process and at each stage along the way we are left with the energy, material, space-time, and biological residues, just as the first instants of the big bang can be observed today, billions of years after the event, thanks to the study of the cosmic microwave background. Cosmologists are the explorers of the universe in the same way that archeologists explore the origins of human societies. Could the evolution of biological systems be intimately linked to the local evolution of the geometry of space-time and its quantum fluctuations, and might these form the backdrop of the different biological mutations that we observe? So many questions! Bit by bit, however, quantum cosmology is starting to shed some light on the subject.

In order to properly assess the qualitative and quantitative dimensions of life and the manifestation of consciousness that it supports, we still have to reflect on the nature of a possible *hyper*–space-time that would include the four-dimensional space-time of general relativity. This hyper–space-time would only be partially revealed in certain dimensions according to the different perspectives or brains of the observers. Four-dimensional space is only the space of representation in which we confine certain aspects of the universe such as movement and evolution in a dichotomized vision of the world: I the ego on one hand and the material universe—the outside world—on the other.

What then of the space-time that I apply to the place and time in which I find myself as a conscious being with material and emotional complexity? Four dimensions should be enough to reveal this complexity. But can my emotions be represented in it? Do they operate in the space-time of general relativity or in the space-time of a superior overarching dimension?

Are four dimensions enough to account for my energy, my physionomy, my consciousness, my emotions, and my intellectual capacities, not to mention my movements and my evolution? Clearly not. Then, in which hyperspace are these categories of being manifested? Are they the same from one species to another and at the different phases of evolution of any given hyperspace? The Qur'anic texts refer to the possibility of these "worlds" belonging to each creature. They would not be interpenetrable, though they may converge at certain points and in certain circumstances. I am not suggesting that this is a plausible explanation but rather that this is an invitation to reflect and construct explanatory models.

This intimation of the foundational text kindles the scientific intuition that needs to go to the next level up in the construction of working hypotheses and mathematical formalization before embarking in the observations at the cosmic level or in laboratory experimentation. There are levels of consciousness (that are also linked to stages in the evolution of consciousness) just as there are levels of reading of the Qur'anic texts. Paradoxically, though, it is not about prying open the secrets of the text (they are not openable anyway) but about emerging with a vision of the world and a mathematically and technologically operational tool to have access to the world and its interpretation (and not to its conquest and destruction). The attempt was successful at a certain phase of the evolution of Islamic thought but was aborted for reasons internal and external to the Islamic world, which would not be appropriate to discuss here.

I suppose then that Being in its beingness (Heidegger's Dasein) constitutes, with the space-time of our experience, a space-time that has more than four dimensions, and which is itself part of a hyper–space-time. Each subspace-time within this hyper–space-time is manifested by the emergence of an observer or of a consciousness that looks upon it and submits it to experience, and the accessibility of which is limited to the type of consciousness that is manifested.

Hence, the subspace-times are separated by neither hierarchy nor privilege; they merely have the potential to emerge ontologically, endowed with ontology as a founding principle. Even though they inherit something in common, the dimensions of one remain inaccessible to the others, just as the observations that are made in one of the subspaces are only partially accessible to others. Each species, in this way, evolves in a subspace, the dimensions of which dictate the nature of its structure and faculties.

In this representational schema it does not make sense to speak of chronology between the different spaces in question, and the question of the origin in time becomes meaningless as it is far removed from the one that is dictated by our daily experience of time, even from the description we give it in the context of general relativity. What matters is their interconnectivity and the processes that link them together. Even in the simple-case "unique" universe (i.e., our universe), as is described by the big bang theory, the question of origins is not well elucidated or else is avoided altogether. An origin of the universe in time must explain the origin of time itself as it is a part of it: this raises the whole mystery of time itself. How can time have an origin and yet at the same time serve to date the origin of space-time out of which it emerged? There is certainly a link to be elucidated between the reflexivity that characterizes intelligence and the nature of time (which is not necessarily the time of our clocks).

Quest for Meaning

If this is the case and if, besides, the theory of quantum cosmology takes us in the direction of a "multiverse," where should we start in ordering the different universes and what meaning should be ascribed to the process of emergence of these multiple universes? Is there a chronological order for the apparition of these universes or do they emerge simultaneously?

Whatever the case may be, we need to establish whether or not there is a causal link between them and the meaning that should be given to time (Does the emergence of these spaces occur in time? Does time exist between the universes?). Could we posit that universes emerge here and there at random without any apparent causal link? What then links them—if there is a link—and how is it that certain events that happen in one are also accessible in another? Is it a case of the contingence of the laws or the necessity for one to prevail over the other to give coherence and meaning to the laws that manifest themselves within it?

Does information manifest itself in the same way between the universes as it does within each universe? The question can be asked in exactly the same way in the field of paleoanthropology. Without calling into question the Darwinian model of evolution, the diversity of the *homo genera* cannot be corroborated by one and only ancestral origin. All chronology between the species of *homo* becomes meaningless as the emergence of species should be investigated within the context of the dynamic processes that underpin cosmic evolution at the macroscopic scale and their links with the evolution of biological structures at the microscopic level, not merely at the level of the filiation between species and their adaptation to the environment.

In this exhilarating period of scientific information, what is most needed is the emergence of new paradigms, notably, a paradigm on the question of origin and our conception of time. Today the classical paradigms—which have served their cause well—constitute a hindrance in the reading and interpretation of these new scientific findings and the perspectives they reveal for a new vision of the world in which the geometry of the universe, its material and energy content, and the emergence of life and consciousness are no longer seen as entities that are foreign from one another, rather, they are encapsulated within a non-exclusive and coherent vision in which each entity—even when it is expressed as a singularity—is supported and fed by the other entities.

Twenty-first-century science can no longer stay confined within the limits of radical positivism, which has marked science during the past three centuries. Quantum mechanics has led the way by introducing the experimenter in the experimental process, particle physics has raised the issue of the fine-tuning of the universe, and, for its part, non-quantum cosmology has raised the issue of the anthropic principle. The theory of information has asked the question of the physical content of information and the mechanisms of the brain in relation to our psychic humor. In light of this, can we go on being content with a science that reads the world through the lens of causality, reducing it to a set of physical laws that are forever set in stone?

I find it hard to imagine the necessity behind the emergence of the universe and the complexity of its structures without the necessity of a consciousness that the event should occur. Necessity is the response of the needs of a consciousness that defines its very own needs and finalizes them. If there is necessity, it exists to finalize a project and implies the setting into motion of a process. It is not necessary to have stars and galaxies for the universe to exist. Likewise, it is not necessary for a telescope to exist for stars and galaxies to exist. But why is it that stars and galaxies are necessary to justify the project of an observer to find him- or herself looking through the eyepiece of a telescope? Would the universe be as we know it without a conscious observer to observe it? Is it observed by other conscious observers and, if the answer is yes, how does this change the conditions of our existence?

It seems to me that necessity is the corollary of consciousness that becomes conscious of itself, of its environment, and of fate. Awareness of one's self causes the necessity of phenomena—and their causality—to give meaning to the world and to "safeguard" the continuity of self: it is what is expressed by causality. But is the emergence of the world the result of a necessity? What then would be the nature of the conscious act that had felt its necessity? The need to be, to become, to share?

Does the emergence of the world and all that emerges have a significance, or is the quest for meaning merely the need of a sentient being: the need to identify links that will help him to find out who he is or what the world is? Where does this desire to ascribe meaning to things come from if they don't have significance ontologically speaking? Do I discover the significance of things or do I give them significance? What mechanisms are used by my brain to allow me to identify the process, to follow it, and to be satisfied with it? Is this capacity to discover or to give meaning to things—or both—evidence of or the result of free will (or pure chance) that governs the process of evolution?

Besides, when I attribute meaning to things, am I ascribing to them a quality that they lacked beforehand? And in so doing, do I participate in the act of their creation or am I merely attributing to them a tailor-made identity that is reduced to my needs and that comforts me in who I am?

What distinguishes us as humans (at least, what we think distinguishes us) is that we are not concerned by only the physical world and its laws. In a peculiar way our existence revolves around our spiritual values and the value-added production of culture. Our survival as a species is not only determined by the biological processes of natural selection but also on conscious processes such as the production of art, science and technology, morality, and religious values. However, religion occupies a particular position in our social lives. It is not purely a human product, as it comes about as the result of divine will and is communicated through the intermediary of a chosen human being who, according to prophecy and revelation, functions outside the standard norms of the rest of humanity.

Hence, mutations are not just genetic but are caused also by the mediation of prophecy and revelation, which in turn have caused mutations to occur within human societies (and continue to do so). And whether we like it or not, these have changed the face of the history of humanity. The religious phenomenon that escapes all forms of rationality and any attempt to rationalize it stumbles across that which is fundamentally human, namely, the need to give significance to the world and the need to secure our cultural and emotional identity through ties with a similar group. This cannot be explained by the laws of physics and molecular arrangements in the way that we understand them in the context of the prevailing scientific paradigm, but rather belongs to a field of investigation from which we cannot turn our attention.

What then does the emergence of a revelation (or a scientific discovery) correspond to? What is it that causes isolated or small groups of individuals in a given time and space to acquire a new vision of the world, new cultural values, and social links in a way that has the power to transform the destinies of entire peoples? It remains something of a mystery how people such as Jesus, the Prophet Muhammad, or Mahatma Gandhi (at another level) change the face of human history. Part of the explanation can be found in the historical and political circumstances of the time, but they don't explain everything, just as it would be simplistic to describe their actions and impact in terms of their personal psychologies. But what is obvious is that they were all motivated by a quest for self and an ideal over and above the level of human corruption.

A revelation (or a discovery) can be described as the unexpected that comes to give meaning to the world when nothing could have anticipated it, though its need was felt. It is a precise and unexpected response to an unformulated request. But who provoked the response and who is responding?

Once again we are in the presence of an inward look that resources itself from the space-time that lends itself to observation and finds a response in the dynamic that is offered to it by history. But if consciousness in the world is captured by each creature who beholds the light of the world in his perspective, it does not follow that each creature can change the world. Is space-time aware that it is being observed? Does it need to be observed? To what end does it allow itself to be discovered by certain very precise beings, adding the value of mystery?

It occurs to me that we homo sapiens and the universe are subject to a necessity and a need that are not easily reconcilable for those who advocate a dichotomized vision of the world. We reveal ourselves like a necessity to a creator who gave meaning to the world; at the same time, we only see meaning in the world and in our existence in our relation with the creator. Are we participating in an act of creation, the purpose of which is only revealed in knowledge and scientific progress, or are we the result of chance and is the only finality of our actions to respond to our needs here and now, evaporating into thin air as our satisfactions are fulfilled?

In the same way, do the different revelations that humanity has borne witness to and the different religions that have grown out of them belong to the same time and space and do they refer to the same quest? The fact is that even if we admit that the quest is the same from one religion to another, this quest is progressive and has stirred human groups at very different times in their evolution or maturation. The religious phenomenon claims to be universal and cosmic but remains very specific, and we tend to observe it among well-compartmentalized human groups. This raises the question of intercommunitarian and interreligious reconciliation and its limitations, and the price to pay for its realization. In the long term, the reconciliation will only be able to be made in recognition of difference, and the dream of seeing humanity live under the flag of a single religion has totalitarian overtones and could not be achieved without significant shedding of blood.

In any case, as far as I am concerned, communitarian reconciliation is not a declaration of intent or the simple adoption of a position but an act that assumes all its significance if it is carried out wholeheartedly by transforming each and every one's perspective on the other as well as his representation of the world. It is first by looking to the other and intercepting his gaze that we can make a step toward each other. Otherwise, each remains entrenched in the idealized realm of his own universe and in the necessary universality of religious values. The perspective of the other is also the integration of his or her values without the fear of losing ours, so long as the perspectives cross and have a reciprocal exchange.

The experience of Ibn Arabi remains an example of unity and benevolence for any individual on a quest of unity with the divine and the dialogue with the other who finds himself on the opposite shore. According to him, "converted" to the divine "way" thanks to a meeting with Jesus himself, he is one of the fervent representatives of the Muhammadan "way" (*Tariqa Muhammadia*) and one of the symbols of the enlightenment of Islam. Ibn Arabi remains a symbol of unity (*Tawhid*) of faiths, the unity of being and of the world. But the men and women who have put themselves to the service of the divine do not have a voice on Earth. The Earth is collapsing under the feet of those who have put the divine at the service of their power, against the dignity of the poor.

In my experience as a Muslim and in my quest for self, I have experienced the companionship and the perspective of my friends and family but also the expression of peace and the word of justice brought to us by Jesus. With the divine word (*A'issa Kalam Allah*) in the space-time of a Muslim who is attached to the revelation by the Word of the Qur'anic message (Muhammad), what reading and what significance should be given to this invitation to a revelation within another, in the space and time of a being caught in his own Dasein? Is it an invitation to liberate the various revealed messages from the clutches of religion, a form of religion that has secularized them and stripped them of their primary essence, the meeting between the divine and the human?

Perhaps science, in freeing itself from its mechanicism, will be able to open more universal dimensions and perspectives on the future of humanity, and all religions, in giving up the privileged access that they think they have to the divine will, better serve humanity as all needs are human needs and all necessities are human necessities. Up until now the God in whose name so many wars and so much misery have been wreaked has been the invention of those whose only interests have been chaos and wars and the human submission that they result in. Does the divine need us to defend his cause when we have failed him in our alliance?

NOTES

1. The mosaic of images on the Hubble telescope Web site is a good illustration of this. See http://www.stsci.edu.

PART VI

Synthesis

14

Science and Transcendence

Limits of Language and
Common Sense

MICHAEL HELLER

We all are realists. If we were not, the surrounding world would soon destroy us. We must take seriously information given us by our senses. If, when crossing the street, we looked for extrasensory inspiration instead of watching the traffic lights, we would have been very quickly eliminated from this game. Poets and philosophers seem odd and impracticable to others because abstract worlds of ideas divert their sight from earthly things. From our everyday contacts with the surrounding world (but also from many slips and bruises), our common sense is born; that is, the set of practical rules that tell us how to behave in order to minimize the damage the world could inflict upon us.

We like to quote science to justify our common sense. The scientific method is but a sharpening of our common sense. Experience constitutes the basis of every science, and the measuring instruments we use in our laboratories are "prolongations" of our senses. The world of technology, from the computer on my desk to artificial satellites, testifies to the ability of our common sense, which has so efficiently conquered the world of matter. Such views, although flattering and sounding nice to our ears, are totally false. Widely spread imaginations concerning science do not match what science really is. Contemporary physics, this most advanced of all sciences, provides us an example that fatally destroys these imaginations.

What could be more in agreement with our common sense than the fact that we cannot go back to our childhood? Time is irreversible. It flows irrevocably from the past to the future. However, this view of time is not that obvious in physics. We know that to every elementary particle there corresponds an antiparticle. Such an antiparticle has the same mass as the corresponding particle but the opposite electric charge. When a particle collides with its antiparticle, they both change into energy. These are the experimental facts, for the first information about the existence of antiparticles came from theory. Since 1926, it has been known that the motion of an electron is described by the Schrödinger equation. The discovery of this equation by Schrödinger was a major breakthrough. Together with the works of Heisenberg, it has created the foundations of modern quantum mechanics. However, the Schrödinger equation had a serious drawback: it did not take into account the laws of special relativity discovered by Einstein two decades earlier.

Einstein's theory is a physical theory of space and time. Although we can ignore it when dealing with the first approximation of the real world, if we want to be more precise in our investigation of the world, we cannot avoid using a relativistic approach. The relativistic counterpart of Schrödinger's equation was discovered by Dirac in 1928. It turned out that Dirac's equation admitted two types of solutions. One of these types described well the elementary particles known at the time. The remaining solutions referred to similar particles but going back in time. How should this be understood? Dirac was audacious enough to claim that such particles really existed and coined the name "antiparticles." This step was not an easy one. Our common sense had to be put upside down. To make this step easier, Dirac used his imagination to picture the void with holes in it, and he interpreted these holes as antiparticles. It doesn't matter whether we would prefer holes in the void or time flowing backward; our common sense is jeopardized.

Let us consider another example. An atom emits two photons

(quanta of light). They travel in two different directions and, after a certain lapse of time, they are far away from each other (it does not matter how far; they can even be at two opposite edges of the galaxy). Photons have the property called "spin" by physicists. It can be measured, and quantum mechanics teaches us that the results of the measurements can assume only two values. Let us denote them symbolically by +1 and −1. However, the situation is much more delicate than our inert language allows us to express. Strictly speaking, we cannot claim that an electron possesses the spin in such a manner as we say that Mr. Smith is tall or has twenty dollars in his pocket. When we are measuring the photon's spin, it behaves as if it were always there. In fact, before the act of measurement, the photon had no spin. Before the act of measurement, a probability existed that the act of measurement would yield, if performed, a given result with a given probability. Let us assume that we have performed the measurement obtaining the result +1. In such a case, on the strength of the laws of quantum mechanics, another photon acquires spin −1, even if it is at the other edge of the galaxy. How does this photon *instantaneously* know about our measurement on the first photon and the result it yields?

This experiment was invented as a purely *Gedanken* experiment by Einstein, Podolsky, and Rosen in 1935 in order to show that the laws of quantum mechanics lead to nonsensical conclusions. However, the physicists—against the opinion of Einstein and his two collaborators— were not much surprised when Allain Aspect, together with his team, performed Einstein's *Gedanken* experiment in reality, and it has turned out that quantum mechanics was right. Aspect was able to perform this experiment owing to enormous progress in experimental methods but also owing to a theoretical idea of John Bell that enabled him to express Einstein's intuitions in the form of precise formulae (the so-called Bell inequalities), which could be compared with the results of measurements.

What happens to photons in Aspect's experiments? When our intuition fails, we must look for help from the mathematical structure of the theory. In quantum mechanics, two photons that once interacted with each other are described by the same vector of state. Strictly speaking, positions of elementary particles behave like spin; an elementary particle is nowhere in space until its position is measured. The state vector of a given quantum object contains information only about probabilities of outcomes of various measurements.

We are met here not only with particles that live "backward in time" but also with particles for which space distances are no obstacles. It looks as if elementary particles did not exist in space and time—as if space and time were only our macroscopic concepts, the usual meaning of which breaks down as soon as we try to apply them to the quantum world. Moreover, can one speak about the individuality of a particle (before its properties are measured) that exists neither in space nor in time? If we agree to consider as a single object something that is described by a single vector of state, could we treat two photons (which previously interacted with each other) situated at two different edges of the galaxy as the single object?

Contemporary physics has questioned the very applicability to the quantum world of such fundamental concepts as space, time, and individuality. Is not our common sense put upside down?

Some philosophers claim that what cannot be said clearly is meaningless. The intention of this claim is praiseworthy; its aim is to eliminate verbosity, which does not contain any substance. However, modern physics has taught us that the possibilities of our language are limited. There are domains of reality—such as the quantum world—at the borders of which our language breaks down. This does not mean that within such domains anything goes—far from it. It turns out that mathematics constitutes a much more powerful language than our everyday means of communication. Moreover, mathematics is not only a language that describes what is seen by our senses but also is a tool that discloses those regions of reality that without its help would forever remain inaccessible for us. All interpretational problems of modern physics can be reduced to the following question: How can all these things that are disclosed by the mathematical method be translated into our ordinary language?

I think that the greatest discovery of modern physics is that our common sense is limited to the narrow domain of our everyday experience. Beyond this domain a region extends to which our senses have no access.

Schrödinger's Question

The world of classical mechanics seemed simple and obvious but, in fact, it never was simple or obvious. The methods discovered by Galileo and Newton did not consist in performing many experiments with pendulums and freely falling bodies, the result of which would later be described with the help of mathematical formulae. Newton, led by his genius, posed a few hazardous hypotheses that suggested to him the mathematical shape of the laws of motion and those of universal gravity. His formulae did not describe the results of experiments. Nobody ever saw a particle uniformly moving to infinity because it was not acted upon by any forces. Moreover, there is no such particle in the entire universe. And it is exactly this statement that is at the very foundation of modern mechanics.

The world of classical mechanics is doubtlessly richer than the world we penetrate with our senses. The most fundamental principle of physics was discovered within the domain of classical mechanics—a principle that could be reached only by mathematical analysis. It is called the *principle of the least action* and its claim is indeed extraordinary. It asserts that every physical theory—from classical mechanics to the most modern quantum field theory—can be constructed in the same way. First, one must correctly guess a function called Lagrangian (which is different for different theories). Then, one computes an integral of this function, called action. And finally, one obtains the laws of this theory by postulating that the action assumes the extreme value (usually the least one but sometimes the greatest one). Physicists often speak about a superunification of physics, that is, about such a theory that would contain everything in itself. We do not yet have such a theory, but the chances that we will are becoming greater and greater. In fact, we already have, in a sense, the unification of the method; all major physical theories are obtainable from the principle of the least action.

With our senses we cannot grasp the fact that all bodies around us move in such a way that a certain simple mathematical expression (the action) assumes the minimal value. But the bodies move in this way. We live surrounded by things that cannot be seen, or heard, or touched. It was Schrödinger who once asked himself: Which achievements of science have best helped the religious outlook of the world? In his answer to this question, he pointed to the results of Boltzmann and Einstein concerning the nature of time. Time, which can change its direction depending on the fluctuations of entropy—which can flow differently in different systems of reference—is no longer a tyrant Chronos, whose absolute regime destroys all our hopes for nontemporal existence, but a physical quantity with a limited region of applicability. If Schrödinger lived today, he could add many new items to his list of achievements that teach us the sense of mystery. Personally, I think, however, that particular scientific achievements do not do this work best but rather the scientific method itself does so. Spectacular results of the most recent physical theories are but examples of what was present in physics for a long time, although it was only understood by a very few.

Two Experiences of Humankind

If we pause for a moment in our competition for new achievements to look backward on the progress of science during the last two centuries, we can see an interesting regularity. In the nineteenth century, humankind went through the great experience of the efficiency of the scientific method. It was a deep experience. Today, we speak of the century of "vapor and electricity" with a touch of irony in our voice. We must know, however, that the road from a candle to the electric bulb, and from a horse-drawn carriage to the railroad train, was longer and more laborious than that from the propeller plane to the intercontinental jet. In the twentieth century, technology made a great jump, but in the nineteenth century, it had started almost from nothing. Yet even then it was obvious that it would change the shape of the civilized world. In the nineteenth century, technology was treated, like never before or after, as a synonym of progress and of the approaching new era of overwhelming happiness. Positivistic philosophy, regarding science as the only valuable source of knowledge, and scientism, wanting to replace philosophy and religion with science, could be considered a philosophical articulation of this great experience—the experience of the efficiency of the scientific method. In the nineteenth century, any suggestion that there were any limits beyond which the scientific method did not work would have been regarded as a senseless heresy. Nobody would have taken it seriously.

The nineteenth century came together with its wars and revolutions. In my opinion, the revolution that took place in the foundations of physics, in the first decades of the twentieth century (and which, I think, is still taking place), had more permanent results for our culture than the political turmoil that shaped the profile of our times. First of all, it turned out that classical mechanics—once believed to be the theory of everything—has, in fact, but a limited field of applicability. It is limited on two sides: from below—in the domain of atoms and elementary particles, where the Newtonian laws must be replaced by the laws of quantum mechanics; and from above—for objects moving with a speed comparable to that of light, where classical physics breaks down and should be replaced by Einstein's theory of relativity. Moreover, the new theories are also, in a sense, limited: the finite value of Planck's constant essentially limits the questions that can be asked in quantum physics, and the finite velocity of light determines horizons of the information transfer in the theory of relativity and cosmology.

The method of physics used from Galilean and Newtonian times (and possibly even from the time of Archimedes) consists in applying mathematics to the investigation of the world. The certainty of mathematical deductions is transferred to physics, which is one of the two sources of the efficiency of the physical method (the other one being controlled experiment). It came as a shock when, in the third decade of the twentieth century, Kurt Gödel proved his famous theorems, which assert that limitations are inherent in mathematics itself: no system of axioms could be formulated from which entire mathematics could be deduced (or even a part of mathematics that is at least as rich as arithmetic). Such a system would be either incomplete or self-contradictory.

Today, there is no doubt that the twentieth century has confronted us with the new great experience—the experience of limitations inherent in the scientific method. Philosophers have understood this relatively late. In the first half of the twentieth century, positivism, in its radical form of logical empiricism, dominated the scene. Only in the 1960s did it become evident that one cannot philosophically support an outdated vision of science. I do not here have in mind those antiscientific and anti-intellectual currents that nowadays so often fanatically fight science in the name of supposed interests of humanity. I have in mind a philosophy of science that recognizes the epistemological beauty of science and its rational applications in the service of man, but does this based on the correct evaluation of both scientific method and the limitations inherent in it.

Science and Transcendence

Science can be compared to a great circle. The points in its interior denote all scientific achievements. What is outside the circle represents not yet discovered regions. Consequently, the circumference of the circle should be interpreted as a place in which what we know today meets with what is still unknown, that is, as a set of scientific questions and unsolved problems. As science progresses, the set of achievements increases and the circle expands; but, together with the area inside the circle, the number of unanswered questions and unsolved problems becomes bigger and bigger. It is historical truth that each resolved problem poses new questions calling for new solutions.

If we agree to understand the term *transcendence*—as suggested by its etymology—as "something that goes beyond," then what is outside the circle of scientific achievements is transcendent with respect to what is inside it. We can see that transcendence admits a graduation: something may go beyond the limits of this particular theory, or beyond the limits of all scientific theories known until now, or beyond the limits of the scientific method as such. Do such ultimate limits exist?

Usually three domains are quoted as forever inaccessible to all attempts of the mathematico-empirical method: the domain of existence, the domain of ultimate rationality, and the domain of meaning and value.

How does one justify the existence of the world? Why does something exist rather than nothing? Some more optimistic physicists believe that in the foreseeable future one will be able to create the unique theory of everything. Such a theory would not only explain everything but would also be the only possible theory of that type. In this way, the entire universe would be understood; there would be no further questions. Let us suppose that we have such a theory—the set of equations fully describing (modeling) the universe. One problem would remain: How can one change from the abstract equations to the real world? What is the origin of those existents that are described by the equations? Who or what ignited the mathematical formulae with existence?

Science investigates the world in a rational way. Knowledge is rational if it is rationally justified. Here, new questions arise: Why should we rationally justify our convictions? Why is the strategy of rational justifications so efficient in investigating the world? One cannot give a rationally justified answer to the first of these questions. Let us try doing this; that is, let us try to rationally justify the statement that everything should be rationally justified. However, our justification (our proof) cannot presuppose what it is supposed to justify (to prove). Therefore, we cannot assume that our convictions should be rationally justified. Consequently, when constructing our proof we cannot use rational means of proving (because they presuppose that we are to prove something); that is, the proof cannot be carried out.

There is no other way out of this dilemma but to assume that the postulate to rationally justify our convictions is but our *choice*. We have two options, and we must choose one of them: when doing science, we either do it in a rational way or we admit an irrational way of doing science. Rationality is a *value*. This can be easily seen if rationality is confronted with irrationality. We evaluate rationality as something good and irrationality as something bad. When choosing rationality we choose something good. It is, therefore, a moral choice. The conclusion cannot be avoided; at the very basis of science there is a moral option. This option was made by humankind when it first formulated questions addressed to the world and started to look for rationally justified answers to them. The entire subsequent history of science could be regarded as a confirmation of this option.

Now follows the second question: Why is the strategy of rational justifications so efficient in studying the world? One could risk the following answer: The fact that our rational methods of studying the world lead to such wonderful results suggests that our choice of rationality is somehow consonant with the structure of the world. The world is not a chaos but an ordered rationality—or, the rational method of science turns out to be so efficient because the world is permeated with meaning.

We should not understand this in an anthropomorphic manner. Meaning, in this context, is not something connected with the human consciousness; it is this property of the world because of which the world discloses its ordered structure, provided it is investigated with the help of rational methods.

Schrödinger's Question Once More

After all these considerations, it would be worthwhile to go back to Schrödinger's question: Which achievements of science have best helped the religious outlook of the world? I think that contemporary science teaches us, as never before, the sense of mystery. In science, we are confronted with mystery at every step. Only outsiders and mediocre scientists believe that in science everything is clear and obvious. Every good scientist knows that he or she is dancing on the edge of a precipice between what is known and what is only feebly felt in just-formulated questions. They also know that the newly born questions open vistas that go beyond the possibilities of our present imagination—imagination that has learned its art in contact with these pieces that we so painfully extracted from the mysteries of the world.

Let us imagine a very good scientist of the nineteenth century—for instance, Maxwell or Boltzmann—who is informed about recent developments of general relativity or quantum mechanics by his younger colleague coming to him from our twenty-first century. Maxwell or Boltzmann would never believe in such "nonsense." Now consider this question: How would we behave if a physicist from the twenty-second century told us about his textbook physics? Only a very shortsighted scientist can be unaware of the fact that he is surrounded by mysteries.

Of course, I have in mind relative mysteries, that is, such mysteries as now go beyond the limits of our knowledge but perhaps tomorrow will become well-digested truths. Do not such mysteries point toward the Mystery (with the capital M)? Does not what today transcends the limits of science suggest something that transcends the limits of all scientific methods? I have expressed these ideas in the form of questions on purpose. Plain assertions are too rigid; they assert something that is expressed by words and syntactic connection between them but remain silent about what is outside the linguistic stuff. Therefore, let us stick to questions that open our intuition to regions not constrained by grammatical rules.

- Are these unimaginable achievements of science, which revolutionize our vision of the world (time flowing backward, cured space-time, particles losing their individuality but communicating with each other with no interaction of space and time), not clear suggestions that reality is not exhausted in what can be seen, heard, touched, measured, and weighed?
- Does not the fact that there exists something rather than nothing excite our metaphysical anxiety?
- Does the fact that the world is not only an abstract structure—never a written formula, an equation solved by nobody, yet something that can be seen, heard, touched, measured, and weighed—direct our thought to the Ultimate Source of Existence?

- Does not the fact that the world can, after all, be put into abstract formulae and equations suggest to us that the abstract thought is more significant than concrete matter?
- Does the rationality that is presupposed but never explained by every scientific investigation not express a reflection of the rational plan hidden in every scientific question addressed to the universe?
- Does not the moral choice of the rationality that underlies all science offer a sign of the Good that is in the background of every correct decision?

These questions are not situated far away, "beyond the limits." The concreteness of existence, the rationality of the laws of nature, the meaning touched by us when we make our decisions are present in every atom, in every quantum of energy, in every living cell, in every fiber of our brain.

It is true that the Mystery is not in the theorems of science but in its horizon. Yet this horizon permeates everything.

Contributors

KHALIL CHAMCHAM, currently carrying out his research at the University of Oxford, earned his first Ph.D. in nuclear physics from the University Claude Bernard, Lyon, France, in 1983, and his second Ph.D. in astrophysics from the University of Sussex, UK, in 1995, and he also holds an MSt in science and religion. His main research interests are star formation, cosmology, Christology, and scientific development in the Islamic world.

PHILIP CLAYTON is Ingraham Professor at the Claremont School of Theology and professor of philosophy and religion at the Claremont Graduate University. His is the author or editor of some fourteen books, including *The Problem of God in Modern Thought; God and Contemporary Science; In Whom We Live and Move and Have Our Being;* and *Mind and Emergence: From Quantum to Consciousness. The Oxford Handbook of Religion and Science,* which he edited, will be published in 2006.

RAMANATH COWSIK has held faculty positions at the University of California at Berkeley and the Tata Institute of Fundamental Research in India. Currently, he is Vainu Bappu Distinguished Professor at the Indian Institute of Astrophysics, Bangalore, and professor of physics at the McDonnell Center for the Space Sciences of Washington University, St. Louis. In 1972, Cowsik pioneered the idea that weakly interacting particle relicts created in the early moments of the big bang will gravitationally dominate the universe, trigger the formation of large-scale structures, and thus generically form halos of dark matter enveloping galaxies. He has been elected Fellow of the Indian Academies of Sciences and as a Foreign Associate of the National Academy of Sciences (U.S.).

PAUL DAVIES is a physicist and cosmologist who is currently professor of natural philosophy at the Australian Centre for Astrobiology at Macquarie University, Sydney. His research ranges from the origin of the universe to the origin of life, topics on which he has published several hundred papers and articles. He is also the author of twenty-six books. Several of them, including *The Mind of God*, address the deeper implications of discoveries at the forefront of scientific research. Davies is the recipient of the 1995 Templeton Prize, as well as several international awards in recognition of his research and communication skills.

CHRISTIAN DE DUVE shared the 1974 Nobel Prize in Physiology or Medicine for his pioneering work on cell structure and function, and has devoted his career to studying the biochemistry of life. A native of Belgium, he has reached emeritus status at the Catholic University of Louvain, Belgium, his alma mater, and the Rockefeller University in New York. Dr. de Duve is the founder of the International Institute of Cellular and Molecular Pathology in Brussels, where he served as its president director from 1974 to 1991 and is now on its board of directors. He is also a member of numerous academies and learned societies, including the U.S. National Academy of Sciences, the American Philosophical Society, and the Royal Society.

BERNARD D'ESPAGNAT is an emeritus professor at the University of Paris, Orsay, and was a research physicist at the French National Center for Scientific Research, at the Fermi National Accelerator Laboratory in Chicago, at the Niels Bohr Institute in Copenhagen, and at CERN in Geneva. In 1959, d'Espagnat joined the University of Paris, where he was professor at both the Paris and Orsay campuses. Professor d'Espagnat was director of the Laboratoire de Physique Theorique et Particules Elementaires, Orsay, from 1970 to 1987. In 1996, he was elected into the Institut de France (Académie des Sciences morales et politiques) as a philosopher of science.

BRUNO GUIDERDONI is an astrophysicist at the French National Center for Scientific Research (CNRS) and a world expert in galaxy formation. He now serves as the director of the observatory in Lyon. He also leads the Islamic Institute for Advanced Studies, and participates in international programs that explore the interface between science and religion.

MICHAEL HELLER, a Roman Catholic priest, is professor of philosophy at the Pontifical Academy of Theology in Cracow, Poland, and is an adjunct member of the Vatican Observatory staff. He twice held the Lemaître Chair at the Catholic University of Louvain, Belgium. Father Heller is an ordinary member of the Pontifical Academy of Sciences in Rome, a founding member of the International Society for Science and Religion, a member of several other international societies, and has written more than twenty books.

JEAN KOVALEVSKY, an astronomer currently at the Observatoire de la Côte d'Azur in Grasse, worked at the Paris Observatory (Bureau des longitudes in Paris) from 1955–74. His main fields are celestial mechanics, space geodesy, metrology, and astrometry using essentially space techniques. He is a member of the French Academy of Sciences and, at present, is involved in the preparation of the European Space Agency space astrometry program, GAIA.

THIERRY MAGNIN is professor of solid-state physics at L'Ecole Nationale Supérieure des Mines, St. Étienne, France, and head of the Material Science Research Laboratory (URA) at the Centre National de la Recherche Scientifique (French National Center for Scientific Research [CNRS]). He has written two hundred papers and reviews as well as five books on solid-state physics, and won the prestigious Laureat Award of the French Academy of Sciences in 1991. Professor Magnin is a Catholic priest and General Vicar of the Diocese of St. Étienne.

THOMAS ODHIAMBO attended Makerere University College and Cambridge University and trained as an entomologist. He was the first professor to head the department of entomology at the University of Nairobi and was the first dean of the faculty of agriculture. In addition, Odhiambo was the founding director of the International Center of Insect Physiology and Ecology (ICIPE) in 1967, was the first president of the African Academy of Sciences, and was the recipient of the Africa Prize in 1987.

WILLIAM D. PHILLIPS is an atomic, molecular, and optical physicist and received a Ph.D. in physics from the Massachusetts Institute of Technology in 1976. After two years as a postdoc at MIT, he joined the National Institute of Standards and Technology (formerly the National Bureau of Standards) in Gaithersburg, Maryland. He is currently the leader of NIST's Laser Cooling and Trapping Group and is also a member of the faculty of the University of Maryland. In 1997, he shared the Nobel Prize in Physics "for development of methods to cool and trap atoms with laser light."

JEAN STAUNE is founder and general secretary of the Interdisciplinary University of Paris, assistant professor of philosophy of science in the MBA course of L'École des Hautes Études Commmerciales (HEC) in Paris, editor of the Temps des Sciences series published by Fayard (Hachette Group), and a member of the European Society for the Study of Science and Theology (ESSSAT).

TRINH XUAN THUAN is professor of astronomy at the University of Virginia. He currently teaches a course in astronomy for nonscientists. His primary research involves studying very young dwarf galaxies in the local universe; Thuan recently co-authored a study that identifies what is possibly the youngest known galaxy within this universe. He has written many articles on the synthesis of primordial elements, galaxy formation, and evolution. He is also the author of several books for the general public, including *The Secret Melody, Chaos and Harmony,* and *The Quantum and the Lotus.*

CHARLES H. TOWNES, a physicist and astronomer, was born in Greenville, South Carolina, in 1915, has BA and BS degrees from Furman University, and a Ph.D. from California Institute of Technology. He has taught at Columbia University, Massachusetts Institute of Technology, and the University of California, served as provost of MIT, and advised the U.S. government on scientific and technical matters. His Nobel Prize in Physics is for initiation of the maser and laser.

AHMED ZEWAIL is the Linus Pauling Chair Professor of Chemistry and Professor of Physics, the director of the Physical Biology Center for Ultrafast Science and Technology and the NSF Laboratory for Molecular Sciences at the California Institute of Technology. Dr. Zewail was awarded the Nobel Prize in 1999, is the recipient of some thirty honorary degrees, and serves as an elected member of national and international academies and societies around the world.

Index